Developments in Primatology: Progress and Prospects

Series Editor: Louise Barrett

For further volumes:
http://www.springer.com/series/5852

Sindhu Radhakrishna · Michael A. Huffman
Anindya Sinha
Editors

The Macaque Connection

Cooperation and Conflict between Humans and Macaques

Springer

Editors
Sindhu Radhakrishna
National Institute of Advanced Studies
Indian Institute of Science Campus
Bangalore, India

Anindya Sinha
National Institute of Advanced Studies
Indian Institute of Science Campus
Bangalore, India

Nature Conservation Foundation
Gokulam Park, Mysore, India

Michael A. Huffman
Department of Social Systems Evolution
Primate Research Institute
Kyoto University
Inuyama, Aichi, Japan

ISBN 978-1-4614-3966-0 ISBN 978-1-4614-3967-7 (eBook)
DOI 10.1007/978-1-4614-3967-7
Springer New York Heidelberg Dordrecht London

Library of Congress Control Number: 2012942237

Printed on acid-free paper

Springer is part of Springer Science+Business Media (www.springer.com)

If mountains shiver in the cold
with what
will they wrap them?

If space goes naked
with what
shall they clothe it?

Allama Prabhu[1]
12th century AD

[1] Ramanujan AK (1973) Speaking of Siva. Penguin Books, New Delhi, p 151

Preface

The concept of this book arose from a symposium entitled "Human-Macaque Interactions: Traditional and Modern Perspectives on Cooperation and Conflict" that we organized at the 23rd Congress of the International Primatological Society, held in Kyoto in September 2010. The symposium highlighted the many aspects of human-macaque relations that exist today and brought to attention the various forms of interactions between macaques and humans, the changes in human attitudes toward macaques over the ages, and the conservation implications of these changes. Apart from some of the participants of this symposium, we also invited other scholars from this field of work to contribute to this volume. Our goal in editing this volume was to document the myriad facets of the association between humans and macaques over the ages, and we believe this book succeeds in doing this. We are indebted to many friends, colleagues, and family members for seeing this volume through and supporting us in all ways possible, be it sourcing literature, answering questions, or putting up with our preoccupied and distracted minds; they include B K Anitha, Hamsa Kalyani, Yoshi Kawamoto, Vaidehi Herbert, Anil Govind, Kakoli Mukhopadhyay, Hemalatha Radhakrishna, and N Radhakrishna. We are grateful to Mrs Rajeshwari Tejaswi for granting us copyright permission to translate and use the story *Gaadli* by K P Poornachandra Tejaswi. We would also like to extend a special note of thanks to all our reviewers who took time out to read the chapters and offer their constructive and timely suggestions – Etienne Benson, Loretta Ann Cormier, A R Vasavi, Janette Wallis, Latha Raman, T R Shankar Raman, Mewa Singh, and Martha Ann Selby. And last, but not the least, we would like to place on record our deep appreciation for the help and support offered by our editors at Springer – Aiko Hiraguchi, Melissa Higgs, and Janet Slobodien.

Bangalore, India — Sindhu Radhakrishna
Inuyama, Aichi, Japan — Michael A Huffman
Bangalore, India — Anindya Sinha

Contents

Contributors

Neel Ahuja Department of English and Comparative Literature, University of North Carolina-Chapel Hill, Chapel Hill, NC, USA

Debapriyo Chakraborty Primate Programme, Nature Conservation Foundation, Mysore, India

National Centre for Biological Sciences, University of Agricultural Sciences Campus, Bangalore, India

Agustin Fuentes Department of Anthropology, University of Notre Dame, Notre Dame, IN, USA

Michael A. Huffman Department of Social Systems Evolution, Primate Research Institute, Kyoto University, Inuyama, Aichi, Japan

Shanthala Kumar Riverstone Ruby Apartment, Vallalar Nagar, Near Chinmaya Vidyalay, Vadavalli, Coimbatore, Tamil Nadu, India

Honnavalli N. Kumara Conservation Biology, Sálim Ali Centre for Ornithology and Natural History, Coimbatore, TN, India

National Institute of Advanced Studies, IISc Campus, Bengaluru, India

Els van Lavieren Moroccan Primate Conservation Foundation, Randwijk, Amsterdam, The Netherlands

Ann MacLarnon Centre for Research in Evolutionary and Environmental Anthropology, University of Roehampton, London, UK

Bonaventura Majolo School of Psychology, University of Lincoln, Lincoln, UK

Avanti Mallapur Director of Biodiversity Conservation, Department of Biodiversity Conservation, Aaranyaa, DeSoto, TX, USA

Laëtitia Maréchal Centre for Research in Evolutionary and Environmental Anthropology, University of Roehampton, London, UK

Garry Marvin Centre for Research in Evolutionary and Environmental Anthropology, University of Roehampton, London, UK

Matthew R. McLennan Faculty of Humanities and Social Sciences, Department of Social Sciences, Oxford Brookes University, Oxford, UK

Yoshihisa Mito Formerly Japan Monkey Centre, Inuyama, Japan

Kakoli Mukhopadhyay 44/1 Nandidurga Road, Ramya Royal Apartments, Bangalore, India

Charmalie A.D. Nahallage Department of Sociology and Anthropology, University of Sri Jayewardenepura, Gangodawila, Nugegoda, Colombo, Sri Lanka

Jeffrey V. Peterson Department of Anthropology, San Diego State University, San Diego, CA, USA

Nancy E.C. Priston Faculty of Humanities and Social Sciences, Department of Social Sciences, Oxford Brookes University, Oxford, UK

Sindhu Radhakrishna National Institute of Advanced Studies, Indian Institute of Science Campus, Bangalore, India

Erin P. Riley Department of Anthropology, San Diego State University, San Diego, CA, USA

Stuart Semple Centre for Research in Evolutionary and Environmental Anthropology, University of Roehampton, London, UK

Anindya Sinha National Institute of Advanced Studies, Indian Institute of Science Campus, Bangalore, India

Nature Conservation Foundation, Mysore, India

David Glenn Smith Department of Anthropology and California National Primate Research Center, University of California, Davis, CA, USA

David S. Sprague Ecosystem Informatics Division, National Institute for Agro-Environmental Sciences, Tsukuba, Ibaraki, Japan

Mohamed Qarro Ecole Nationale Forestiére d'Ingénieurs, Salé, Hassane Rabat, Morocco

Part I
Introduction

Monkey on monkeyman's stick
puppet at the end of a string

I've played as you've played
I've spoken as you've told me
I've been as you've let me be

Mahadeviyakka[1] *1130 – 1160 CE*

[1]Ramanujan AK (1973) Speaking of Siva. Penguin Books, New Delhi, p 117

Chapter 1
The Gulf Between Men and Monkeys*

Sindhu Radhakrishna

1.1 Introduction

The term human–animal interaction involves a whole gamut of associations, from competition and commensalism to conflict, cooperation, coexistence and companionship, between humans and other animal species. Yet, conflict is the most common frame that defines human–wildlife interactions today (Conover 2002; Woodroffe et al. 2005). Contemporary scientific accounts of human and other animal interactions generally tend to explore the nuances of conflict between the species, the more specific and localised drivers behind the discord, solutions for problem mitigation and end with adjurations for more equitable relations between humans and their co-inhabitants on this planet. The reasons for this are not far to seek. Globally, problems due to wildlife are attracting more attention, and it has been claimed that the intensity of human–wildlife conflict has not only risen in recent times but is likely to escalate further in the future (Madden 2004). Increasingly, nonhuman animal populations are losing out in the competition for space and resources that they face with the human species, and both for utilitarian and utopian reasons, humans need be convinced that they must tolerate and share the available space and resources with their nonhuman brethren.

Although human–wildlife conflict is emphasised a great deal in conservation biology, a notable advance in the discipline (and in related fields) in more recent times is the growing body of work on the ethical and moral imperatives governing human interactions with other animal species. Many biologists and conservationists have argued that animals experience emotions in much the same way that humans do;

*I owe the title of this chapter to Cook (1999) who talks about the gulf between men and monkeys in his essay on Ibn Qutayba and the monkeys.

S. Radhakrishna (✉)
National Institute of Advanced Studies, Indian Institute of Science Campus, Bangalore 560 012, India
e-mail: sindhu@nias.iisc.ernet.in

S. Radhakrishna et al. (eds.), *The Macaque Connection: Cooperation and Conflict between Humans and Macaques*, Developments in Primatology: Progress and Prospects 43, DOI 10.1007/978-1-4614-3967-7_1, © Springer Science+Business Media, LLC 2013

therefore an acknowledgement of animal sentience should primarily regulate the way we interact with other animal species (Bradshaw et al. 2005; Jickling and Paquet 2005; Bekoff 2002, 2006; Dawkins 2006; Goodall 2006; Simmonds 2006). Secondly, though humans share this earth with other animal and plant species, human needs have taken precedence over all else. Human actions have degraded the natural environment to such an extent that the future of many wildlife species stands at risk. Apart from causing the extinction of wildlife populations through exploitation of habitat and resources, human activities have also brought about 'suffering, fear, physical injury, psychological trauma and disease in wild animals' (Bekoff 2002; Bradshaw et al. 2005; Morrison et al. 2007; Darimont et al. 2009). Hence, it has been suggested that it is important to meld the science of wildlife conservation with compassion for other animals and to work towards shared coexistence rather than wildlife conservation in isolation (Vucetich and Nelson 2007; Paquet and Darimont 2010).

In common with other wildlife species, most discussions on human–other primate interactions also appear as conflict-oriented (Patterson and Wallis 2005). Conflict is a critical component of the human–other primate interface, but it is essential to recognise that other forms of associations also exist between human and non-human primate species (Fuentes and Wolfe 2002; Fuentes and Hockings 2010). As a group, macaques are perhaps the best exemplars of the multiplicity of relationships that may exist between human and other animal species. In their interactions with humans, macaques often play multiple roles that transcend the boundaries of categorisation—they are simultaneously pets and symbols of wild nature, deities and crop raiders, marauders and protected wards, employees and companions, surrogate humans and dispensable animals. The curious tendency of many macaque species to be strongly commensal labels them as common and borderland; they are neither fully wild nor domesticated and therefore are problematic units for management and conservation (Richard et al. 1987; Leong 2009; Radhakrishna and Sinha 2011a, b). For these reasons, they are excellent models for investigations into human–animal relations; they act as both 'mirrors and windows' for our understanding of human interactions with non-human species (Mullin 1999; Knight 2005).

1.2 Cooperation and Conflict with Macaques

Conflict and cooperation, it is contended, are integral aspects of societies (Hobbes 1651/1929; Dugatkin 2000). As two ends of a spectrum of associations that humans share with other humans and with non-human species, they are also interesting points to launch enquiries into the nature of the human–macaque interface. Humans interact with macaques in various ways, with the latter serving as pets, food, putative medicinal material, commodities of trade, entertainers, harvesting labourers, crop raiders, wild animals, religious symbols, commensal species, pathogen transmitters, experimental objects and cultural signifiers. Conflict between the two species occurs either when both seek the same resource, but only one can gain it, or when both have different objectives, with one attempting to prevent the other from achieving its goal (Preuschoft and van Schaik 2000). According to this interpretation, crop raiding and property damage by macaques and hunting or trapping macaques for food or other purposes by humans are clearly sources of conflict between the species. However, the nature of the other interactions is not so well defined. As humans make use of macaques as pets, entertainers, labourers and experimental objects, scientists and philosophers concerned with animal rights would label these interactions as conflict-mediated (Regan 1984; Palmer and Sandøe 2011). A biological characterisation of these interactions, on the other hand, would veer towards cooperation.

In biology, a broad definition of cooperation refers to two or more organisms acting together to accomplish some common goal (Price 2011). In the field of human–other animal interactions, documentation of cooperation between humans and wild animals is not unknown, but certainly uncommon. The most well-known example of human–wildlife cooperation is the case of dolphins and fishers working together to obtain fish (Busnel 1973; Pryor et al. 1990; Peterson et al. 2008; Zappes et al. 2011). Describing the practice of using pig-tailed macaques to pick coconuts in Thailand, Sponsel et al. (2002) propose that this is a cooperative relationship between humans and macaques as both species gain from it. Humans profit economically, while the macaques obtain food, shelter and protection from hunting. The authors aver that although the macaques may be unequal partners in that they are not able to exercise their choice or free will in this relationship, cooperation in the strictest biological sense does not require the actions of the involved parties to be *voluntary.* Although the inequity involved in removing male pig-tailed macaque infants from their natal groups in the wild (in order to train them to pick coconuts) is undeniable, it must be noted that oftentimes the relationship between keeper and macaque is similar to that of parent and offspring (Sponsel et al. 2002).

The deep bond that typically develops between humans and the animals they care for, and the behavioural characteristics of the animals that make this possible, has been remarked upon in several contexts but particularly in that of domestication (Campbell 2005; Hart 2005; Marvin 2005). The concept of domestication, with its adherent image of dominant man and subordinate animal, is fertile ground for debates on the intimacy of man–other animal relations (Knight 2005). Rejecting the common and anthropocentric view of domestication that sees humans as solely responsible for removing some animal species from the wild habitat and utilising them for their purposes, some authors have argued that domestication arose as a by-product of a mutualistic or commensal association that existed between humans and certain animals (Reed 1980; Brothwell 1983; O'Connor 1997; Budiansky 1999). Associating with humans benefited some animal species because it reduced food competition from larger species and provided greater protection from predators. Hence, these species may have 'actively sought the closer relationship with humans', and these commensal associations paved the way for domestication (O'Connor 1997). Domestication therefore can be seen as symbiotic or cooperative since both humans and the partner animals benefit from the process (Bokonyi 1989; O'Connor 1997; Budiansky 1999).

By the same analogy, interactions between macaques and humans, wherein both species obtain some benefit from the other, may be categorised as cooperative.

Humans profit from their associations with macaques when they use them as pets, entertainers, labourers and for experimental purposes (see Ahuja, Fuentes, this volume). Although macaques do obtain food and protection in return, it can be argued that wild macaque populations also achieve these ends quite effectively, without the interference of humans. Instead, an alternative proposition to understand how macaques could gain from these interactions is to see these relationships as offshoots of macaque commensalism. O'Connor (1997) suggests that early humans assigned roles such as domestic, companion and helper to various commensal species in a bid to benefit from animals that were clearly 'not going to go away'. It is probable that the human–macaque interrelationship also moved along similar lines. Initially, the macaque could have been merely a commensal species found near human habitations; soon people may have attached cultural import to the species and begun utilising it for its functional aspects. Classical literature from some parts of the world supports such a view. Some of the early writings on human–macaque interactions merely record the commensal nature of the macaque; the religious significance of the macaque which is such a prominent part of the human–macaque interface today appears to be a later phenomenon (see Mito and Sprague, Radhakrishna, this volume). Macaques have certainly benefited from their commensality with humans. For one, their cultural and religious importance ensures proximate gains in terms of food and shelter, as well as long-term survival advantages. More importantly, their geographic spread is unrivalled in the non-human primate world, and until the 1990s at least, species population densities were, for the most part, relatively stable (IUCN 2011).

Gudger (1923) quotes the example of monkeys trained to pick fruits in Egypt around 2500 BCE (as evidenced by rock paintings of this era) to demonstrate that the practice of using monkeys as harvesters is an ancient one that has extended to present times. The author's final remark in this article is thought-provoking, commenting that the concept of using monkeys for human tasks has survived a passage of more than 5,000-odd years; he ends by stating: 'Verily there is nothing new under the sun'. Indeed, if one only considers the generalities of man–other animal association, this would appear to be true. However, some particulars have changed—human population size has increased, land is more of a premium now and cultural values are different—and this has created a formidable transformation in our attitudes towards animals and wildlife.

1.3 Social Construction and the Power of Perception

We call it a grain of sand,
but it calls itself neither grain nor sand.
It does just fine without a name,
whether general, particular,
permanent, passing,
incorrect, or apt.

Szymborska (1995)

It has been argued that the term human–wildlife conflict is a misnomer because it emphasises 'conscious antagonism between wildlife and humans'; in most instances, the issue is one of human–wildlife competition for resources or human–human conflict, as it is actually a disagreement among different human groups over how wildlife and their needs should be regarded or managed (Madden 2004; Peterson et al. 2010). Emphasising the importance of appropriate language and phrasing in discussing problems that arise from human–wildlife interactions, Peterson et al. (2010) contend that the rhetorical framing of such issues critically influences the way people react to the situations in reality. This standpoint of course harks back to the larger debate concerning social construction (Schneider and Ingram 1993; Soule and Lease 1995; Czech et al. 1998; Crist 2004). Social constructivists hold that all our assessments and representations regarding the world around us are mediated through our social contexts; in other words, we ascribe meaning to phenomena or concepts that surround/influence us based as much on our cultural experiences and social learning as on the properties of the subject in question. As noted by several studies, social construction deeply impacts the tone and form of human–other animal interactions, and by extension, the welfare of the animals involved in such relations (Evernden 1992; Czech et al. 1998; Herda-Rapp and Goedeke 2005; Leong 2009). (A related note of philosophical enquiry is of course the much more fundamental question about the perceived distinction between human and animal that was so provocatively raised by Derrida (2008).)

The power of social construction is particularly evident in the varied and shifting modes of human–macaque relations over the ages. One of these has to do with what is seen as the most problematic face of human–macaque interactions, at least as far as humans are concerned, that is, macaque depredations. Expressed as crop raiding in rural areas and damage to property and kitchen gardens in urban areas, conflict due to macaque species and its effective resolution has been the subject of much discussion and heartburn in the recent decades (Patterson and Wallis 2005; Gumert et al. 2011). But damage or danger to human life and property due to wildlife is not a recent or even a modern phenomenon. Classical Tamil poetry from southern India dating back to 100 BCE describes crop raiding by wild animals and birds and people's attempts to guard their fields from them (Hart and Heifetz 1999; Selby 2011). Popular sayings and local legends in many cultures suggest that crop losses due to wildlife were considered acceptable and a natural way of life (Sutlive 1978; Ohnuki-Tierney 1991; Morris 1998, 2000). A crucial difference in modern times may be the lowered tolerance for wildlife crop depredations. Knight (2000) proposes that technological advances in the field of agriculture brought in its wake assurances of higher productivity and protection from typical pests; this has resulted in farmers having higher expectations of crop yields and therefore being less accepting of crop damage.

Apart from a general reduction in forbearance levels, another critical factor that affects people's attitudes towards crop-raiding wildlife has to do with perceived notions regarding the amount of damage caused by a species. Perceptions regarding the destruction wrought by a wildlife species dictate reactions against the species far more strongly than do actual losses (Knight 2000; Gillingham and Lee 2003; Naughton-Treves and Treves 2005). Examinations of macaque crop-raiding occurrences underline the relevance of both these parameters for deeper insights into how social construction shapes the character of human–macaque interfaces (Singh and Rao 2004; Chalise and Johnson 2005; Wang et al. 2006; Marchal and Hill 2009; Riley and Priston 2010). For example, farmers in central Sulawesi consider the Tonkean macaque as most detrimental to their cacao production primarily because the macaques raid the farms even when the farmers are present. Yet quantitative investigations regarding crop losses revealed that in actuality, forest rats cause greater fiscal damage than do macaques (Riley 2007).

Yet another example of the role of social construction in influencing people's reactions towards macaques is the difference in human demeanours towards 'forest macaques' and 'temple macaques'. Across much of south and southeast Asia, significant numbers of macaques can be found inhabiting temple premises (Aggimarangsee and Brockelman 2005; Loudon et al. 2006; Medhi et al. 2007; Nahallage and Huffman, this volume). These macaque populations are typically provisioned and protected from any possible harm and enjoy an elevated status in the minds of the devotees frequenting the temples. However, similar goodwill may not always be extended to the same macaque species outside the boundaries of the temple complexes (Loudon et al. 2006; Malaivijitnond and Hamada 2008; Nahallage and Huffman, this volume). This appears to contradict the tenet of sacredness that is often invoked to explain why macaques are tolerated in regions where Hinduism, Buddhism and Jainism are predominant (Eudey 1994; Strum 1994; Wheatley 1999). Fuentes et al. (2005) suggest that this puzzle resolves itself when it is understood that macaques are protected in temples not because they are sacred, but because they are residents of a 'sacred geography'. The sanctity that is allied to temples and other icons of religious significance such as sacred groves and particular tree species is also passed on to those beings that are found within its precincts. It is for this reason that even bats are revered when they are found within these environments, but not treated with the same indulgence when found outside, or separate from these elements. A similar position is reiterated by Peterson and Riley (this volume) who show that 'perceptions of macaque sacredness are more strongly tied to space than to an inherent holiness for monkeys in the eyes of Balinese Hinduism'.

In his essay on comparing human–other animal relations among hunter-gatherers and in pastoral societies, Ingold (1994) draws attention to a fundamental aspect of human–animal relations, that of characterisation or description. He observes that 'Just as humans have a history of their relations with animals, so also animals have a history of their relations with humans'. The crucial difference is that only humans narrate their history (Ingold 1994), creating a form of asymmetry in the relation. Narratives lend power to a thought, action or deed, and few subjects demonstrate this more forcefully than human–other animal relations. Human narratives, through their depictions of macaques, have strongly affected macaque lives in the past and will, in all probability, continue to do so in the future. A case in question is the recent attempt to declare the rhesus macaque as 'vermin' in the state of Uttarakhand in northern India so that farmers and ordinary people at the receiving end of rhesus macaque depredations can kill animals opportunistically (Radhakrishna and Sinha 2011b). Although humans manifestly have the upper hand in scripting these narratives, it may perhaps be instructive to remember that as subjects of these narratives, macaques are also co-players in this theatre of interactions.

1.4 Macaque Connections

The concept of this book and its title owes much to Bryant (1979). In his elegant paper on recognising the importance of human–other animal relationships, Bryant catalogues the numerous ways in which animals routinely influence our lives and calls upon sociologists to more keenly appreciate this 'zoological dimension'. Much of our own behaviour, he asserts, will be seen through new eyes if we take into account the 'zoological connection' (Bryant 1979). This book is an attempt to capture the essence of those connections that exist between macaques and humans. Our cultural heritage shows that macaques have strongly impacted human lives and thoughts over the ages. In turn humans have also, rather powerfully, modified the course of macaque lives. The more obvious and much discussed effects naturally concern macaque conservation and future survival. As the essays in this volume reveal, in many parts of the world, regard for macaque species is fast being replaced by hostility and anger due to the destruction and damage caused by the animals, and this is serious cause for alarm, not only for wildlife biologists and conservationists, but also for sociologists and humanists. The essays in Part II of this book present snippets of the human gaze vis-à-vis macaques in earlier times and the consequences of those perspectives for macaque lives. Harlow's essay not only recaptures the debate around what is possibly the most infamous experimental use of macaques, it also succeeds in presenting the human perspective behind the laboratory utilisation of monkeys. The human–macaque interface in Japan has always been of interest to primatologists, not only due to the unique beginnings of primatology in Japan, but also because of the unusual philosophy of subjectivity that has shaped the Japanese approach to primatology. In their chapter, Mito and Sprague trace the history of human–macaque interface in Japan and try to explain how people relate to monkeys on this island nation. The last essay in this section on the representations of monkeys in classical Tamil poetry delineates the commensal nature of the macaque more than 2,000 years ago. Radhakrishna points out that in the classical Tamil era, macaques were a part of the natural landscape for people, along with other elements like trees, elephants, wild pigs and birds, and unlike the complexities of present times, a simple acceptance characterised the human–macaque interface then.

The essays in Part III explore various facets of cooperative relations between macaques and humans. Among macaques, the rhesus macaque has always been an important laboratory primate, and many of our biomedical advances stem from successful experiments on the species. Ahuja's paper documents the growing significance of the rhesus in research laboratories in the USA during a certain period in the past, the political machinations surrounding its import and breeding and its embedment as a national symbol in the political history of America. Ahuja posits that the end of colonisation led to a different kind of domestication of the rhesus; captive breeding of the species was now actively encouraged within the USA and the earlier image of the animal as an untamed, wild, colonial species transformed to that of a docile, national, laboratory resident. A different perspective on human–macaque cooperation is offered by Mallapur who examines the use of macaque tourism by humans as a tool for macaque conservation. Tourism tends to boost the economic development of the tourist destination; for this reason, macaque tourism has been recommended as a practical way to engage local human populations in the conservation of macaque species and their habitats. However, this method also has undesirable consequences such as food provisioning and its ramifications on macaque population size, higher possibilities of anthropozoonotic disease transmission and negative effects on macaque social behaviour. Fuentes' paper on macaques as commodities engages with the discourse that macaques are co-participants in a shared ecology. Recommending that we move beyond traditional approaches that study macaque behavioural ecology and identify the human–macaque interface with human uses of macaques or the costs of competition between the species, Fuentes uses the broader framework of multispecies ethnoprimatology to analyse how macaques function as certain kinds of commodities in human societies.

Part IV of this book deals with the most well-known feature of the human–macaque interface, that is, conflict. The contributions to this section comprise a set of case studies (for want of a better word) from various parts of the macaque world and showcase people's reactions to crop-raiding macaques in different regions. The first composition in the section, Gaadli, is a story about crop-raiding macaques in southern India. Comprising all the familiar elements of human–macaque conflict in Asia—poor farmers, macaque sacredness, inefficient mitigation methods and translocation—the story whimsically recounts how people typically react to crop-raiding macaques in the region. Nahallage and Huffman's paper on human–macaque interface in Sri Lanka records people's views on the toque macaque, the various ways in which people interact with the species and the conservation implications of this for the future survival of the species. Peterson and Riley present a snapshot of the macaque situation in Indonesia and report the different aspects of human–macaque interactions in the country. In the second part of their essay, they examine what they term the 'paradox of macaque sacredness'. Many studies have commented on the sacredness of macaques in Indonesia, yet conflict with macaques is also an important part of the human–macaque interface in the country. Peterson and Riley present preliminary results from their study in Sulawesi to show that reverence for macaque is often conditional to the species residence in holy places. The last chapter in this section talks about human–macaque interface with regard to the lone African species, the Barbary macaque. Majolo et al. chronicle the history of human–macaque relations in Morocco, Algeria and Gibraltar and the conservation threats faced by the species due to their conflicts with humans and conclude by presenting insights gained from their study on macaque tourism in Gibraltar.

The final part of this book tackles an often ignored topic—how living beside humans has affected macaque populations. Across Asia, macaques, perhaps more than any other animal species, exemplify the multiple outcomes of synurbisation and the conservation problems of commensal species. With the rapid encroachment of human populations into forested areas, some macaque species have dwindled in number, while others have increasingly moved in greater numbers into human habitats. In the case of the former, the spectre of species extinction looms very large indeed, and in the case of the latter, the urban troops are clearly doomed to a losing battle for survival. A secondary, and largely ignored, consequence has to do with changes in macaque behaviour and society due to human interactions. In the first essay in this section, Sinha and Mukhopadhyay summarise the results of their long-term study on bonnet macaques and offer their understanding of how anthropogenic factors have influenced changes in social structures and relationships within a bonnet macaque population in southern India. The next chapter in this book by Chakraborty and Glenn Smith investigates the genetic consequences of anthropogenic factors on macaque populations across the world. Using the phylogenetic histories of the Barbary macaque, the Japanese macaque and the long-tailed macaque in Mauritius as examples, the authors identify three main axes along which humans have critically impinged macaque biology through their actions. The final chapter in this book by Priston and McLennan completes the enquiry into how living near humans has affected macaques by reviewing the various strategies that are used by humans to manage macaques. The authors first present an overview of the different forms of human–macaque conflict that commonly occur and then discuss the types of mitigation options that are employed by humans to deal with problem macaque populations. They conclude by reminding the readers that all occurrences of human–macaque conflicts are embedded in particular cultural contexts and that resolution mechanisms must bear in mind that a single strategy may not work in all situations.

Certain aspects are missing in this volume: Some macaque species (e.g. the lion-tailed macaque and the stump-tailed macaque) and their interactions with humans find no mention, nor are there any debates on the moral concerns of human treatment of macaques. Although the specific context of human–macaque interaction is unique for every macaque species, a delineation of all the different circumstances that frame human–macaque interactions in different parts of the world would result in a very cumbersome tome. Hence the objective in this volume has been to emphasise some main trajectories rather than to capture every possible feature. However, the absence of a full-bodied discussion on the ethical issues surrounding human–macaque interface may be seen as a more serious lacunae. Due to the wide range of human–macaque interactions, the ethical considerations of human treatment of macaques are rarely simple or straightforward matters. Hence, rather than engage in polemical stances on the morality of human actions concerning macaques, the attempt in this volume has been to highlight various contexts wherein the ethical connotations of human behaviour towards macaques assume considerable importance. For example, Mallapur, Majolo et al. and Sinha and Mukhopadhyay review the animal welfare concerns of using macaques for tourism benefits; Fuentes underlines the moral disquiets of keeping macaques as pets and performers, and many of the other chapters (Nahallage and Huffman, Peterson and Riley, Priston and McLennan) point out the travails borne by macaques as a result of interactions with humans. The most prominent face of the animal ethics debate, namely, the utilisation of macaques for laboratory experiments, is addressed in two different ways by the Harlow and Ahuja chapters. While Harlow's essay showcases some of the viewpoints that fuel the animal experimentation controversy, Ahuja demonstrates how the changing representations of the rhesus macaque in the cultural consciousness of the USA was mediated not only by geopolitical realities but also by scientific pursuits that evidenced the biological links between man and monkeys and underscored the cognitive and sentient capacities of primate species.

This volume thus brings together not only representations of the diverse human–macaque connections that exist but also the divergent views of scholars involved in this area of work. How do people perceive macaques? As the chapters in this volume attest, macaques are sometimes seen as passive laboratory specimens that may be utilised for particular purposes, at other times, as sentient beings, with rich, emotional lives. In some cultures, they are simply elements of the landscape, while in others, they represent the exotic wild. They may fulfil roles as companion individuals with unique personalities; equally they may be icons of sacredness and tourist attractions. To some groups of people, they are conservation-worthy wildlife; to others, annoying pests. While some of the essays are prescriptive in tone, offering suggestions on the way to move forward, others are more reflective, highlighting particular singularities that mark our affiliation with the macaques. The goal behind editing this book was to capture the myriad perspectives that embody human–macaque relationships, and hopefully, not only primatologists and anthropologists but all enthusiasts of animal studies will find that objective fulfilled.

Acknowledgements I am immensely grateful to T R Shankar Raman, Michael Huffman and Anindya Sinha for their patient reading of this chapter; their insightful comments helped to significantly improve an earlier version of this chapter.

References

Aggimarangsee N, Brockelman WY (2005) Monkey-human interactions in Thailand. Am J Phys Anthropol, vol 126 (Suppl 40):62–63

Bekoff M (2002) Minding animals: awareness, emotions, and heart. Oxford University Press, New York

Bekoff M (2006) Animal emotions and animal sentience and why they matter: blending 'science sense' with common sense, compassion and heart. In: Turner J, D'Silva J (eds) Animals, ethics and trade: the challenge of animal sentience. Earthscan, London, pp 27–40

Bokonyi S (1989) Definitions of animal domestication. In: Clutton-Brock TH (ed) The walking larder: patterns of domestication, pastoralism, and predation. Unwin Hyman, London, pp 22–27

Bradshaw GA, Schore AN, Brown JL, Poole JH, Moss CJ (2005) Elephant breakdown: social trauma: early disruption of attachment can affect the physiology, behaviour and culture of animals and humans over generations. Nature 433:807

Brothwell DR (1983) Why on earth the guinea-pig? The problem of restricted mammal exploitation in the new world. In: Proudfoot B (ed) Site, environment and economy, vol 73, International series. British Archaeological Reports, Oxford, pp 115–119

Bryant CD (1979) The zoological connection: animal-related human behavior. Soc Forces 58: 399–421

Budiansky S (1999) The covenant of the wild: why animals chose domestication. Yale University Press, New Haven

Busnel RG (1973) Symbiotic relationship between man and dolphins. Trans N Y Acad Sci 35: 112–131

Campbell B (2005) On 'loving your water buffalo more than your own mother': relationships of animal and human care in Nepal. In: Knight J (ed) Animals in person: cultural perspectives on human-animal intimacy. Berg, Oxford, pp 79–100

Chalise MK, Johnson RL (2005) Farmer attitudes toward the conservation of "pest" monkeys: the view from Nepal. In: Paterson JD, Wallis J (eds) Commensalism and conflict: the human-primate interface. American Society of Primatologists Publications, Oklahoma, pp 223–239

Conover MR (2002) Resolving human-wildlife conflicts: the science of wildlife damage management. Lewis Publishers, Boca Raton

Crist E (2004) Against the social construction of nature and wilderness. Environ Ethics 26:5–24

Czech B, Krausman PR, Borkhataria R (1998) Social construction, political power and the allocation of benefits to endangered species. Conserv Biol 12:1103–1112

Darimont CT, Carlson SM, Kinnison MT, Paquet PC, Reimchen TE, Wilmers CC (2009) Human predators outpace other agents of trait change. Proc Natl Acad Sci 106:952–954

Dawkins MS (2006) Through animal eyes: what behaviour tells us. Appl Anim Behav Sci 100:4–10

Derrida J (2008) The animal that therefore I am. Fordham University Press, New York

Dugatkin L (2000) Cheating monkeys and citizen bees: the nature of cooperation in animals and human beings. Harvard University Press, Cambridge

Eudey AA (1994) Temple and pet primates in Thailand. Revue d'Ecologie (La Terre et la Vie) 49:273–280

Evernden N (1992) The social creation of nature. Johns Hopkins University Press, Baltimore

Fuentes A, Hockings K (2010) The ethnoprimatological approach in primatology. Am J Primatol 72:841–847

Fuentes A, Wolfe LD (eds) (2002) Primates face to face: the conservation implications of human-nonhuman primate interface. Cambridge University Press, Cambridge

Fuentes A, Southern M, Suaryan KG (2005) Monkey forests and human landscapes: is extensive sympatry sustainable for *Homo sapiens* and *Macaca fascicularis* in Bali. In: Patterson JD, Wallis J (eds) Commensalism and conflict: the primate-human interface. American Society of Primatology Publications, Oklahoma, pp 168–195

Gillingham S, Lee P (2003) People and protected areas: a study of local perceptions of wildlife crop-damage conflict in an area bordering the selous game reserve, Tanzania. Oryx 37:316–325

Goodall J (2006) The sentience of chimpanzees and other animals. In: Turner J, D'Silva J (eds) Animals, ethics and trade: the challenge of animal sentience. Earthscan, London, pp 3–11

Gudger EW (1923) Monkeys trained as harvesters: instances of a practice extending from remote times to the present. Nat Hist 23:272–279

Gumert MD, Fuentes A, Jones-Engel L (2011) Monkeys on the edge: ecology and management of long-tailed macaques and their interface with humans. Cambridge University Press, Cambridge

Hart LA (2005) The elephant-mahout relationship in India and Nepal: a tourist attraction. In: Knight J (ed) Animals in person: cultural perspectives on human-animal intimacy. Berg, Oxford, pp 161–190

Hart G L, Heifetz H (trans ed) (1999) The Purananuru: four hundred songs of war and wisdom. Columbia University Press, New York, pp 15–37

Herda-Rapp A, Goedeke TL (2005) Mad about wildlife: looking at social conflict over wildlife. Brill, Boston

Hobbes T (1651/1929) Leviathan, or the matter, forme and power of a common-wealth ecclesiasticall and civil. Clarendon, Oxford
Ingold T (1994) From trust to domination: an alternative history of human-animal relations. In: Manning A, Serpell J (eds) Animals and human society: changing perspectives. Routledge, London, pp 1–22
IUCN (2011) IUCN red list of threatened species. Version 2011.2. www.iucnredlist.org
Jickling R, Paquet PC (2005) Wolf stories: reflections on science, ethics, and epistemology. Environ Ethics 2:115–134
Knight J (2000) Introduction. In: Knight J (ed) Natural enemies: people-wildlife conflicts in anthropological perspective. Routledge, London, pp 1–35
Knight J (2005) Introduction. In: Knight J (ed) Animals in person: cultural perspectives on human-animal intimacy. Berg, Oxford, pp 1–14
Leong KM (2009) The tragedy of becoming common: landscape change and perceptions of wildlife. Soc Nat Resour 23:111–127
Loudon JE, Howells ME, Fuentes A (2006) The importance of integrative anthropology: a preliminary investigation employing primatological and cultural anthropological data collection methods in assessing human-monkey co-existence in Bali, Indonesia. Ecol Environ Anthropol 2:2–13
Madden F (2004) Creating coexistence between humans and wildlife: global perspectives on local efforts to address human–wildlife conflict. Hum Dimens Wildl 9:247–257
Malaivijitnond S, Hamada Y (2008) Current situation and status of long-tailed macaques (*Macaca fascicularis*) in Thailand. Nat Hist J Chulalongkorn Univ 8:185–204
Marchal V, Hill C (2009) Primate crop-raiding: a study of local perceptions in our villages in north Sumatra, Indonesia. Primate Conserv 24:107–116
Marvin G (2005) Disciplined affections: the making of an english pack of foxhounds. In: Knight J (ed) Animals in person: cultural perspectives on human-animal intimacy. Berg, Oxford, pp 61–78
Medhi R, Chetry D, Basavdatta C, Bhattacharjee PC (2007) Status and diversity of temple primates in northeast India. Primate Conserv 22:135–138
Morris B (1998) The power of animals. Berg, Oxford
Morris B (2000) Animals and ancestors: an ethnography. Berg, Oxford
Morrison JC, Sechrest W, Dinerstein E, Wilcove DS, Lamoreux JF (2007) Persistence of large mammal faunas as indicators of global human impacts. J Mammal 88:1363–1380
Mullin MH (1999) Mirrors and windows: sociocultural studies of human-animal relationships. Annu Rev Anthropol 28:201–224
Naughton-Treves L, Treves A (2005) Socio-ecological factors shaping local support for wildlife: crop-raiding by elephants and other wildlife in Africa. In: Thirgood S, Rabinowitz A (eds) People and wildlife: conflict or coexistence. Cambridge University Press, Cambridge, pp 252–277
O'Connor TP (1997) Working at relationships: another look at animal domestication. Antiquity 71:149–156
Ohnuki-Tierney E (1991) Embedding and transforming polytrope: the monkey as self in Japanese culture. In: Fernandez JW (ed) Beyond metaphor: trope theory in anthropology. Stanford University Press, Stanford, pp 159–189
Palmer C, Sandøe P (2011) Animal ethics. In: Appleby MC, Mench JA, Olsson IAS, Hughes BO (eds) Animal welfare, 2nd edn. CAB International, Oxfordshire, pp 1–12
Paquet PC, Darimont CT (2010) Wildlife conservation and animal welfare: two sides of the same coin? Anim Welf 19:177–190
Patterson J, Wallis J (eds) (2005) Commensalism and conflict: the primate–human interface. American Society of Primatologists Publications, Oklahoma
Peterson D, Hanazaki N, Simões-Lopes PC (2008) Natural resource appropriation in cooperative artisanal fishing between fishermen and dolphins (*Tursiops truncatus*) in Laguna, Brazil. Ocean Coast Manag 5:469–475
Peterson ML, Birckhead JL, Leong K, Peterson MJ, Peterson TR (2010) Rearticulating the myth of human–wildlife conflict. Conserv Lett 3:74–82

Preuschoft S, van Schaik CP (2000) Dominance and communication: conflict management in various social settings. In: Aureli F, de Waal FBM (eds) Natural conflict resolution. University of California Press, Berkeley, pp 77–105
Price ME (2011) Cooperation as a classic problem in behavioural biology. In: Swami V (ed) Evolutionary psychology: a critical introduction. Wiley-Blackwell, New York, pp 73–106
Pryor K, Lindbergh J, Lindbergh S, Milano R (1990) A dolphin-human fishing cooperative in Brazil. Mar Mamm Sci 6:77–82
Radhakrishna S, Sinha A (2011a) Less than wild: commensal primates and wildlife conservation. J Biosci 36:749–753
Radhakrishna S, Sinha A (2011b) Dr Jekyll and Mr Hyde: the strange case of human-macaque interactions in India. Curr Conserv 4:39–40
Reed CA (1980) The beginnings of animal domestication. In: Cole HH, Garrett WN (eds) Animal agriculture: the biology, husbandry, and use of domestic animals. W. H. Freeman, San Francisco, pp 3–20
Regan T (1984) The case for animal rights. Routledge, London
Richard AF, Goldstein SJ, Dewar RE (1987) Weed macaques: the evolutionary implications of macaque feeding ecology. Int J Primatol 10:569–594
Riley EP (2007) The human-macaque interface: conservation implications of current and future overlap and conflict in Lore Lindu National Park, Sulawesi, Indonesia. Am Anthropol 109: 473–484
Riley EP, Priston NEC (2010) Macaques in farms and folklore: exploring the human–nonhuman primate interface in Sulawesi, Indonesia. Am J Primatol 71:1–7
Schneider AL, Ingram I (1993) Social construction of target populations: implications for politics and policy. Am Polit Sci Rev 87:334–347
Selby MA (2011) Tamil love poetry: the five hundred short poems of the Ainkurunuru. Columbia University Press, New York
Simmonds MP (2006) Into the brains of whales. Appl Anim Behav Sci 100:103–116
Singh M, Rao N (2004) Population dynamics and conservation of commensal bonnet macaques. Int J Primatol 25:847–859
Soule ME, Lease G (1995) Reinventing nature? Responses to postmodern deconstruction. Island Press, Washington, DC
Sponsel LE, Ruttanadakul N, Natadecha-Sponsel P (2002) Monkey business? The conservation implications of macaque ethnoprimatology in southern Thailand. In: Fuentes A, Wolfe LD (eds) Primates face to face: the conservation implications of human-nonhuman primate interface. Cambridge University Press, Cambridge, pp 288–309
Strum SC (1994) Prospects for management of primate pests. Revue d'Ecologie (La Terre et la Vie) 49:295–306
Sutlive VH Jr (1978) The Iban of Sarawak. AHM Publishing Corporation, Arlington Heights
Szymborska W (1995) View with a grain of sand: selected poems (trans: Baraficzak S, Cavanagh C). Harcourt Brace and Company, Orlando
Vucetich J, Nelson MP (2007) What are 60 warblers worth? Killing in the name conservation? Oikos 116:1267–1278
Wang SW, Curtis PD, Lassoie JP (2006) Farmer perceptions of crop damage by wildlife in Jigme Singye Wangchuck National Park, Bhutan. Wildl Soc Bull 34:359–365
Wheatley BP (1999) The sacred monkeys of Bali. Waveland Press, Prospect Heights
Woodroffe R, Thirgood S, Rabinowitz A (eds) (2005) People and wildlife: conflict or coexistence? Cambridge University Press, Cambridge
Zappes CA, Andriolo A, Simões-Lopes PC, Di Beneditto APM (2011) Human-dolphin (*Tursiops truncatus* Montagu, 1821) cooperative fishery and its influence on cast net fishing activities in Barra de Imbé/Tramandaí, Southern Brazil. Ocean Coast Manag 54:427–432

Part II
Traditional Views of Macaques

There is the princeling macaque, an artful creature,
With a form coarse in appearance and foul manners.

His face is like an old gaffer,
His body resembles a child.

Wang Yanshou[1] *118–138 CE*

[1]Harper D (2001) Poets and primates: Wang Yanshou's poem on the Macaque. Asia Major (Third Ser) XIV:1–25

Chapter 2
The Nature of Love

Harry F. Harlow

Love is a wondrous state: deep, tender, and rewarding. Because of its intimate and personal nature, it is regarded by some as an improper topic for experimental research. But whatever our personal feelings may be, our assigned mission as psychologists is to analyze all facets of human and animal behavior into their component variables. So far as love or affection is concerned, psychologists have failed in this mission. The little we know about love does not transcend simple observation, and the little we write about it has been written better by poets and novelists. But of greater concern is the fact that psychologists tend to give progressively less attention to a motive which pervades our entire lives. Psychologists, at least psychologists who write textbooks, not only show no interest in the origin and development of love or affection, but they seem to be unaware of its very existence.

The apparent repression of love by modem psychologists stands in sharp contrast with the attitude taken by many famous and normal people. The word "love" has the highest reference frequency of any word cited in Bartlett's book of Familiar Quotations. It would appear that this emotion has long had a vast interest and fascination for human beings, regardless of the attitude taken by psychologists, but the quotations cited, even by famous and normal people, have a mundane redundancy. These authors and authorities have stolen love from the child and infant and made it the exclusive property of the adolescent and adult.

This chapter is an excerpt from a paper published by Harlow in American Psychologist (1958, Vol 13), based on his Presidential address at the Sixty-sixth Annual Convention of the American Psychological Association, Washington, D.C., August 31, 1958. Only some of the figures present in the original article have been reproduced here.

S. Radhakrishna et al. (eds.), *The Macaque Connection: Cooperation and Conflict between Humans and Macaques*, Developments in Primatology: Progress and Prospects 43, DOI 10.1007/978-1-4614-3967-7_2,

Thoughtful men, and probably all women, have speculated on the nature of love. From the developmental point of view, the general plan is quite clear: The initial love responses of the human being are those made by the infant to the mother or some mother surrogate. From this intimate attachment of the child to the mother, multiple learned and generalized affectional responses are formed.

Unfortunately, beyond these simple facts, we know little about the fundamental variables underlying the formation of affectional responses and little about the mechanisms through which the love of the infant for the mother develops into the multifaceted response patterns characterizing love or affection in the adult. Because of the dearth of experimentation, theories about the fundamental nature of affection have evolved at the level of observation, intuition, and discerning guesswork, whether these have been proposed by psychologists, sociologists, anthropologists, physicians, or psychoanalysts.

The position commonly held by psychologists and sociologists is quite clear: The basic motives are, for the most part, the primary drives – particularly hunger, thirst, elimination, pain, and sex – and all other motives, including love or affection, are derived or the secondary drives. The mother is associated with the reduction of the primary drives – particularly hunger, thirst, and pain – and through learning, affection or love is derived.

It is entirely reasonable to believe that the mother, through association with food, may become a secondary reinforcing agent, but this is an inadequate mechanism to account for the persistence of the infant-maternal ties. There is a spate of researches on the formation of secondary reinforcers to hunger and thirst reduction. There can be no question that almost any external stimulus can become a secondary reinforcer if properly associated with tissue-need reduction, but the fact remains that this redundant literature demonstrates unequivocally that such derived drives suffer relatively rapid experimental extinction. Contrariwise, human affection does not extinguish when the mother ceases to have intimate association with the drives in question. Instead, the affectional ties to the mother show a lifelong, unrelenting persistence and, even more surprising, widely expanding generality.

Oddly enough, one of the few psychologists who took a position counter to modern psychological dogma was John B. Watson, who believed that love was an innate emotion elicited by cutaneous stimulation of the erogenous zones. But experimental psychologists, with their peculiar propensity to discover facts that are not true, brushed this theory aside by demonstrating that the human neonate had no differentiable emotions, and they established a fundamental psychological law that prophets are without honor in their own profession.

The psychoanalysts have concerned themselves with the problem of the nature of the development of love in the neonate and infant, using ill and aging human beings as subjects. They have discovered the overwhelming importance of the breast and related this to the oral erotic tendencies developed at an age preceding their subjects' memories. Their theories range from a belief that the infant has an innate need to achieve and suckle at the breast to beliefs not unlike commonly accepted psychological theories. There are exceptions, as seen in the recent writings of John Bowlby, who attributes importance not only to food and thirst satisfaction but also to "primary object-clinging," a need for intimate physical contact, which is initially associated with the mother.

As far as I know, there exists no direct experimental analysis of the relative importance of the stimulus variables determining the affectional or love responses in the neonatal and infant primate. Unfortunately, the human neonate is a limited experimental subject for such researches because of his inadequate motor capabilities. By the time the human infant's motor responses can be precisely measured, the antecedent determining conditions cannot be defined, having been lost in a jumble and jungle of confounded variables.

Many of these difficulties can be resolved by the use of the neonatal and infant macaque monkey as the subject for the analysis of basic affectional variables. It is possible to make precise measurements in this primate beginning at two to ten days of age, depending upon the maturational status of the individual animal at birth. The macaque infant differs from the human infant in that the monkey is more mature at birth and grows more rapidly, but the basic responses relating to affection, including nursing, contact, clinging, and even visual and auditory exploration, exhibit no fundamental differences in the two species. Even the development of perception, fear, frustration, and learning capability follows very similar sequences in rhesus monkeys and human children.

Three years' experimentation before we started our studies on affection gave us experience with the neonatal monkey. We had separated more than 60 of these animals from their mothers 6 to 12 hours after birth and suckled them on tiny bottles. The infant mortality was only a small fraction of what would have obtained had we let the monkey mothers raise their infants. Our bottle-fed babies were healthier and heavier than monkey-mother-reared infants. We know that we are better monkey mothers than are real monkey mothers, thanks to synthetic diets, vitamins, iron extracts, penicillin, chloromycetin, 5 % glucose, and constant, tender, loving care.

During the course of these studies, we noticed that the laboratory-raised babies showed strong attachment to the cloth pads (folded gauze diapers) which were used to cover the hardware-cloth floors of their cages. The infants clung to these pads and engaged in violent temper tantrums when the pads were removed and replaced for sanitary reasons. Such contact-need or responsiveness had been reported previously by Gertrude van Wagenen for the monkey and by Thomas McCulloch and George Haslerud for the chimpanzee and is reminiscent of the devotion often exhibited by human infants to their pillows, blankets, and soft, cuddly stuffed toys. Responsiveness by the one-day-old infant monkey to the cloth pad is shown in Figure 1, and an unusual and strong attachment of a six-month-old infant to the cloth pad is illustrated in Figure 2. The baby, human or monkey, if it is to survive, must clutch at more than a straw.

We had also discovered during some allied observational studies that a baby monkey raised on a bare wire-mesh cage floor survives with difficulty, if at all, during the first five days of life. If a wire-mesh cone is introduced, the baby does better; and, if the cone is covered with terry cloth, husky, healthy, happy babies evolve. It takes more than a baby and a box to make a normal monkey. We were impressed by the possibility that, above and beyond the bubbling fountain of breast or bottle, contact comfort might be a very important variable in the development of the infant's affection for the mother.

At this point, we decided to study the development of affectional responses of neonatal and infant monkeys to an artificial, inanimate mother, and so we built a surrogate mother which we hoped and believed would be a good surrogate mother. In devising this surrogate mother, we were dependent neither upon the capriciousness of evolutionary processes nor upon mutations produced by chance radioactive fallout. Instead, we designed the mother surrogate in terms of modern human-engineering principles (Figure 3). We produced a perfectly proportioned, streamlined body stripped of unnecessary bulges and appendices. Redundancy in the surrogate mother's system was avoided by reducing the number of breasts from two to one and placing this unibreast in an upper-thoracic, sagittal position, thus maximizing the natural and known perceptual-motor capabilities of the infant operator. The surrogate was made from a block of wood, covered with sponge rubber, and sheathed in tan cotton terry cloth. A light bulb behind her radiated heat. The result was a mother, soft, warm, and tender, a mother with infinite patience, a mother available twenty-four hours a day, a mother that never scolded her infant and never struck or bit her baby in anger. Furthermore, we designed a mother machine with maximal maintenance efficiency since failure of any system or function could be resolved by the simple substitution of black boxes and new component parts. It is our opinion that we engineered a very superior monkey mother, although this position is not held universally by the monkey fathers.

Before beginning our initial experiment, we also designed and constructed a second mother surrogate, a surrogate in which we deliberately built less than the maximal capability for contact comfort. This surrogate mother is illustrated in Figure 4. She is made of wire-mesh, a substance entirely adequate to provide postural support and nursing capability, and she is warmed by radiant heat. Her body differs in no essential way from that of the cloth mother surrogate other than in the quality of the contact comfort which she can supply. In our initial experiment, the dual mother-surrogate condition, a cloth mother and a wire mother were placed in different cubicles attached to the infant's living cage as shown in Figure 4. For four newborn monkeys, the cloth mother lactated and the wire mother did not; and for the other four, this condition was reversed. In either condition, the infant received all its milk through the mother surrogate as soon as it was able to maintain itself in this way, a capability achieved within two or three days except in the case of very immature infants. Supplementary feedings were given until the milk intake from the mother surrogate was adequate. Thus, the experiment was designed as a test of the relative importance of the variables of contact comfort and nursing comfort. During the first 14 days of life, the monkey's cage floor was covered with a heating pad wrapped in a folded gauze diaper, and thereafter the cage floor was bare. The infants were always free to leave the heating pad or cage floor to contact either mother, and the time spent on the surrogate mothers was automatically recorded. Figure 5 shows the total time spent on the cloth and wire mothers under the two conditions of feeding.

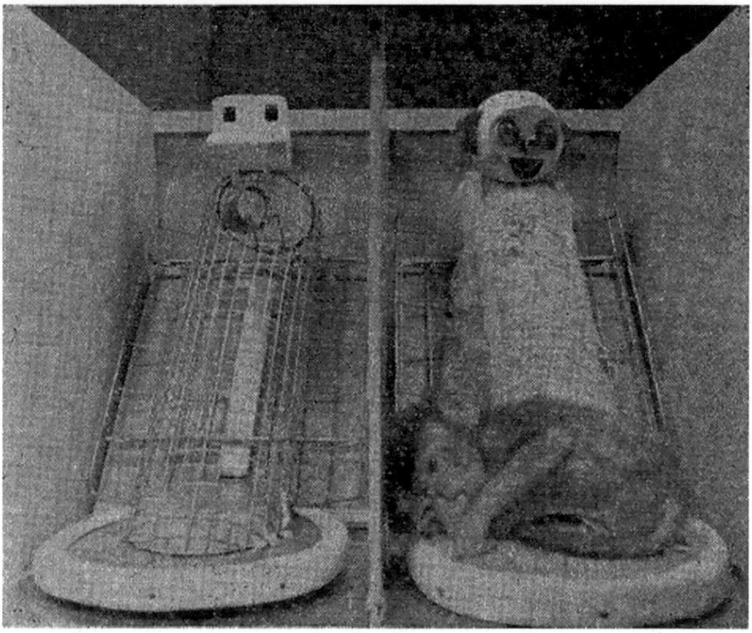

Figure 4 Wire and cloth mother surrogates

These data make it obvious that contact comfort is a variable of overwhelming importance in the development of affectional responses, whereas lactation is a variable of negligible importance. With age and opportunity to learn, subjects with the lactating wire mother showed decreasing responsiveness to her and increasing responsiveness to the nonlactating cloth mother, a finding completely contrary to any interpretation of derived drive in which the mother form becomes conditioned to hunger-thirst reduction. The persistence of these differential responses throughout 165 consecutive days of testing is evident in Figure 6.

One control group of neonatal monkeys was raised on a single wire mother, and a second control group was raised on a single cloth mother. There were no differences between these groups in amount of milk ingested or in weight gain. The only difference between the groups lays in the composition of the feces, the softer stools of the wire-mother infants suggesting psychosomatic involvement. The wire mother is biologically adequate but psychologically inept.

We were not surprised to discover that contact comfort was an important basic affectional or love variable, but we did not expect it to overshadow so completely the variable of nursing; indeed, the disparity is so great as to suggest that the primary function of nursing as an affectional variable is that of insuring frequent and intimate body contact of the infant with the mother. Certainly, man cannot live by milk alone. Love is an emotion that does not need to be bottle- or spoon-fed, and we may be sure that there is nothing to be gained by giving lip service to love.

A charming lady once heard me describe these experiments and, when I subsequently talked to her, her face brightened with sudden insight: "Now I know what's wrong with me," she said, "I'm just a wire mother." Perhaps she was lucky. She might have been a wire wife.

****

One function of the real mother, human or subhuman, and presumably of a mother surrogate is to provide a haven of safety for the infant in times of fear and danger. The frightened or ailing child clings to its mother, not its father, and this selective responsiveness in times of distress, disturbance, or danger may be used as a measure of the strength of affectional bonds. We have tested this kind of differential responsiveness by presenting to the infants in their cages, in the presence of the two mothers, various fear-producing stimuli such as the moving toy bear illustrated in Figure 13. A typical response to a fear stimulus is shown in Figure 14, and the data on differential responsiveness are presented in Figure 15. It is apparent that the cloth mother is highly preferred over the wire one, and this differential selectivity is enhanced by age and experience. In this situation, the variable of nursing appears to be of absolutely no importance: the infant consistently seeks the soft mother surrogate regardless of nursing condition.

Similarly, the mother or mother surrogate provides its young with a source of security, and this role or function is seen with special clarity when mother and child are in a strange situation. At the present time, we have completed tests for this relationship on four of our eight baby monkeys assigned to the dual mother-surrogate condition by introducing them for three min into the strange environment of a room measuring six feet by six feet by six feet (also called the "open-field test") and containing multiple stimuli known to elicit curiosity-manipulatory responses in baby monkeys. The subjects were placed in this situation twice a week for eight weeks with no mother surrogate present during alternate sessions and the cloth mother during the others. A cloth diaper was always available as one of the stimuli throughout all sessions. After one or two adaptation sessions, the infants always rushed to the mother surrogate when she was present and clutched her, rubbed their bodies against her, and frequently manipulated her body and face. After a few additional sessions, the infants began to use the mother surrogate as a source of security, a base of operations. As shown in Figures 16 and 17, they would explore and manipulate a stimulus and then return to the mother before adventuring again into the strange new world. The behavior of these infants was quite different when the mother was absent from the room. Frequently, they would freeze in a crouched position, as is illustrated in Figures 18 and 19. Emotionality indices such as vocalization, crouching, rocking, and sucking increased sharply, as shown in Figure 20. Total emotionality score was cut in half when the mother was present. In the absence of the mother, some of the experimental monkeys would rush to the center of the room where the mother was customarily placed and then run rapidly from object to object, screaming and crying all the while. Continuous, frantic clutching of their bodies was very common, even when not in the crouching position. These monkeys frequently contacted and clutched the cloth diaper, but this action never pacified them. The same

Figure 14 Typical response to cloth mother surrogate in fear test

behavior occurred in the presence of the wire mother. No difference between the cloth-mother-fed and wire-mother-fed infants was demonstrated under either condition. Four control infants never raised with a mother surrogate showed the same emotionality scores when the mother was absent as the experimental infants showed in the absence of the mother, but the controls' scores were slightly larger in the presence of the mother surrogate than in her absence. Some years ago, Robert Butler demonstrated that mature monkeys enclosed in a dimly lighted box would open and reopen a door hour after hour for no other reward than that of looking outside the box. We now have data indicating that neonatal monkeys show this same compulsive visual curiosity on their first test day in an adaptation of the Butler apparatus which we call the "love machine," an apparatus designed to measure love. Usually, these tests are begun when the monkey is 10 days of age, but this same persistent visual exploration has been obtained in a three-day-old monkey during the first half-hour of testing. Butler also demonstrated that rhesus monkeys show selectivity in rate and frequency of door-opening to stimuli of differential attractiveness in the visual field outside the box. We have utilized this principle of response selectivity by the monkey to measure strength of affectional responsiveness in our infants in the baby version of the Butler box. The test sequence involves four repetitions of a test battery in which four stimuli – cloth mother, wire mother, infant monkey, and empty box – are presented for a 30-minute period on successive days. The first four subjects in the dual mother-surrogate group were given a single-test sequence at 40 to 50 days of age, depending upon the availability of the apparatus, and only their data are presented. The second set of four subjects is being given repetitive tests to

Figure 19 Response in the open-field test in the absence of the mother surrogate

obtain information relating to the development of visual exploration. The apparatus is illustrated in Figure 21. The data obtained from the first four infants raised with the two mother surrogates are presented in the middle graph of Figure 22 and show approximately equal responding to the cloth mother and another infant monkey, and no greater responsiveness to the wire mother than to an empty box. Again, the results are independent of the kind of mother that lactated, cloth or wire. The same results are found for a control group raised, but not fed, on a single cloth mother; these data appear in the graph on the right. Contrariwise, the graph on the left shows no differential responsiveness to cloth and wire mothers by a second control group, which was not raised on any mother surrogate. We can be certain that not all love is blind.

The first four infant monkeys in the dual mother-surrogate group were separated from their mothers between 165 and 170 days of age and tested for retention during the following 9 days and then at 30-day intervals for six successive months. Affectional retention as measured by the modified Butler box is given in Figure 23. In keeping with the data obtained on adult monkeys by Butler, we find a high rate of responding to any stimulus, even the empty box. But throughout the entire 185-day retention period, there is a consistent and significant difference in response frequency to the cloth mother contrasted with either the wire mother or the empty box, and no consistent difference between wire mother and empty box.

Affectional retention was also tested in the open field during the first 9 days after separation and then at 30-day intervals, and each test condition was run twice at each retention interval. The infant's behavior differed from that observed during the period preceding separation. When the cloth mother was present in the post-separation period, the babies rushed to her, climbed up, clung tightly to her, and rubbed their heads and faces against her body. After this initial embrace and reunion, they played on the mother, including biting and tearing at her cloth cover, but they rarely made any attempt to leave her during the test period, nor did they manipulate or play with the objects in the room, in contrast with their behavior before maternal separation. The only exception was the occasional monkey that left the mother surrogate momentarily, grasped the folded piece of paper (one of the standard stimuli in the field), and brought it quickly back to the mother. It appeared that deprivation had enhanced the tie to the mother and rendered the contact-comfort need so prepotent that need for the mother overwhelmed the exploratory motives during the brief, three-minute test sessions. No change in these behaviors was observed throughout the 185-day period. When the mother was absent from the open field, the behavior of the infants was similar in the initial retention test to that during the preseparation tests, but they tended to show gradual adaptation to the open-field situation with repeated testing and, consequently, a reduction in their emotionality scores.

In the last five retention test periods, an additional test was introduced in which the surrogate mother was placed in the center of the room and covered with a clear Plexiglas box. The monkeys were initially disturbed and frustrated when their explorations and manipulations of the box failed to provide contact with the mother. However, all animals adapted to the situation rather rapidly. Soon they used the box as a place of orientation for exploratory and play behavior, made frequent contacts with the objects in the field, and very often brought these objects to the Plexiglas box. The emotionality index was slightly higher than in the condition of the available cloth mothers, but it in no way approached the emotionality level displayed when the cloth mother was absent. Obviously, the infant monkeys gained emotional security by the presence of the mother even though contact was denied.

Affectional retention has also been measured by tests in which the monkey must unfasten a three-device mechanical puzzle to obtain entrance into a compartment containing the mother surrogate. All the trials are initiated by allowing the infant to go through an unlocked door, and in half the trials, it finds the mother present and in half, an empty compartment. The door is then locked, and a ten-minute test is conducted. In tests given prior to separation from the surrogate mothers, some of the infants had solved this puzzle, and others had failed. The data of Figure 24 show that on the last test before separation, there were no differences in total manipulation under mother-present and mother-absent conditions, but striking differences exist between the two conditions throughout the post-separation test periods. Again, there is no interaction with conditions of feeding.

The overall picture obtained from surveying the retention data is unequivocal. There is little, if any, waning of responsiveness to the mother throughout this five-month period, as indicated by any measure. It becomes perfectly obvious that this affectional bond is highly resistant to forgetting and that it can be retained for very long periods of time by relatively infrequent contact reinforcement. During the next year, retention tests will be conducted at 90-day intervals, and further plans are dependent upon the results obtained. It would appear that affectional responses may show as much resistance to extinction as has been previously demonstrated for learned fears and learned pain, and such data would be in keeping with those of common human observation.

The infant's responses to the mother surrogate in the fear tests, the open-field situation, and the baby Butler box and the responses on the retention tests cannot be described adequately with words. For supplementary information, we turn to the motion picture record. (At this point, a 20-minute film was presented illustrating and supplementing the behaviors described thus far in the address.)

We have already described the group of four control infants that had never lived in the presence of any mother surrogate and had demonstrated no sign of affection or security in the presence of the cloth mothers introduced in test sessions. When these infants reached the age of 250 days, cubicles containing both a cloth mother and a wire mother were attached to their cages. There was no lactation in these mothers, for the monkeys were on a solid-food diet. The initial reaction of the monkeys to the alterations was one of extreme disturbance. All the infants screamed violently and made repeated attempts to escape the cage whenever the door was opened. They kept a maximum distance from the mother surrogates and exhibited a considerable amount of rocking and crouching behavior, indicative of emotionality. Our first thought was that the critical period for the development of maternally directed affection had passed and that these macaque children were doomed to live as affectional orphans. Fortunately, these behaviors continued for only 12 to 48 h and then gradually ebbed, changing from indifference to active contact on, and exploration of, the surrogates. The home-cage behavior of these control monkeys slowly became similar to that of the animals raised with the mother surrogates from birth. Their manipulation and play on the cloth mother became progressively more vigorous to the point of actual mutilation, particularly during the morning after the cloth mother had been given her daily change of terry covering. The control subjects were now actively running to the cloth mother when frightened and had to be coaxed from her to be taken from the cage for formal testing.

Objective evidence of these changing behaviors is given in Figure 25, which plots the amount of time these infants spent on the mother surrogates. Within 10 days, mean contact time is approximately nine hours, and this measure remains relatively constant throughout the next 30 days. Consistent with the results on the subjects reared from birth with dual mothers, these late-adopted infants spent less than one and one-half hours per day in contact with the wire mothers, and this activity level was relatively constant throughout the test sessions. Although the maximum time that the control monkeys spent on the cloth mother was only about half that spent by the original dual mother-surrogate group, we cannot be sure that this discrepancy is a function of differential early experience. The control monkeys were about three months older when the mothers were attached to their cages than the experimental animals had been when their mothers were removed and the retention tests begun. Thus, we do not know what the amount of contact would be for a 250-day-old animal raised from birth with surrogate mothers. Nevertheless, the magnitude of the differences and the fact that the contact-time curves for the mothered-from-birth infants had remained constant for almost 150 days suggest that early experience with the mother is a variable of measurable importance.

The control group has also been tested for differential visual exploration after the introduction of the cloth and wire mothers; these behaviors are plotted in Figure 26.

By the second test session, a high level of exploratory behavior had developed, and the responsiveness to the wire mother and the empty box is significantly greater than that to the cloth mother. This is probably not an artifact since there is every reason to believe that the face of the cloth mother is a fear stimulus to most monkeys that have not had extensive experience with this object during the first 40 to 60 days of life. Within the third test session, a sharp change in trend occurs, and the cloth mother is then more frequently viewed than the wire mother or the blank box; this trend continues during the fourth session, producing a significant preference for the cloth mother.

Before the introduction of the mother surrogate into the home-cage situation, only one of the four control monkeys had ever contacted the cloth mother in the open-field tests. In general, the surrogate mother not only gave the infants no security but also instead appeared to serve as a fear stimulus. The emotionality scores of these control subjects were slightly higher during the mother-present test sessions than during the mother-absent test sessions. These behaviors were changed radically by the fourth post-introduction test approximately 60 days later. In the absence of the cloth mothers, the emotionality index in this fourth test remains near the earlier level, but the score is reduced by half when the mother is present, a result strikingly similar to that found for infants raised with the dual mother surrogates from birth. The control infants now show increasing object exploration and play behavior, and they begin to use the mother as a base of operations, as did the infants raised from birth with the mother surrogates. However, there are still definite differences in the behavior of the two groups. The control infants do not rush directly to the mother and clutch her violently, but instead they go toward, and orient around, her, usually after an initial period during which they frequently show disturbed behavior, exploratory behavior, or both.

That the control monkeys develop affection or love for the cloth mother when she is introduced into the cage at 250 days of age cannot be questioned. There is every reason to believe, however, that this interval of delay depresses the intensity of the affectional response below that of the infant monkeys that were surrogate-mothered from birth onward. In interpreting these data, it is well to remember that the control monkeys had had continuous opportunity to observe and hear other monkeys housed in adjacent cages and that they had had limited opportunity to view and contact surrogate mothers in the test situations, even though they did not exploit the opportunities.

During the last 2 years, we have observed the behavior of two infants raised by their own mothers. Love for the real mother and love for the surrogate mother appear to be very similar. The baby macaque spends many hours a day clinging to its real mother. If away from the mother when frightened, it rushes to her, and in her presence, it shows comfort and composure. As far as we can observe, the infant monkey's affection for the real mother is strong, but no stronger than that of the experimental monkey for the surrogate cloth mother, and the security that the infant gains from the presence of the real mother is no greater than the security it gains from a cloth surrogate. Next year, we hope to put this problem to final, definitive, experimental test. But whether the mother is real or a cloth surrogate, there does develop a deep and abiding bond between mother and child. In one case, it may be the call of the wild, and in the other, the McCall of civilization, but in both cases, there is "togetherness."

In spite of the importance of contact comfort, there is reason to believe that other variables of measurable importance will be discovered. Postural support may be such a variable, and it has been suggested that, when we build arms into the mother surrogate, 10 is the minimal number required to provide adequate child care. Rocking motion may be such a variable, and we are comparing rocking and stationary mother surrogates and inclined planes. The differential responsiveness to cloth mother and cloth-covered inclined plane suggests that clinging as well as contact is an affectional variable of importance. Sounds, particularly natural, maternal sounds, may operate as either unlearned or learned affectional variables. Visual responsiveness may be such a variable, and it is possible that some semblance of visual imprinting may develop in the neonatal monkey. There are indications that this becomes a variable of importance during the course of infancy through some maturational process.

John Bowlby has suggested that there is an affectional variable which he calls "primary object following," characterized by visual and oral search of the mother's face. Our surrogate-mother-raised baby monkeys are at first inattentive to her face, as are human neonates to human mother faces. But by 30 days of age, ever-increasing responsiveness to the mother's face appears – whether through learning, maturation, or both – and we have reason to believe that the face becomes an object of special attention.

Our first surrogate-mother-raised baby had a mother whose head was just a ball of wood since the baby was a month early and we had not had time to design a more esthetic head and face. This baby had contact with the blank-faced mother for 180 days and was then placed with two cloth mothers, one motionless and one rocking, both being endowed with painted, ornamented faces. To our surprise, the animal would compulsively rotate both faces 180° so that it viewed only a round, smooth face and never the painted, ornamented face. Furthermore, it would do this as long as the patience of the experimenter in reorienting the faces persisted. The monkey showed no sign of fear or anxiety, but it showed unlimited persistence. Subsequently it improved its technique, compulsively removing the heads and rolling them into its cage as fast as they were returned. We are intrigued by this observation, and we plan to examine systematically the role of the mother face in the development of infant-monkey affections. Indeed, these observations suggest the need for a series of ethological-type researches on the two-faced female.

Although we have made no attempts thus far to study the generalization of infant-macaque affection or love, the techniques which we have developed offer promise in this uncharted field. Beyond this, there are few if any technical difficulties in studying the affection of the actual, living mother for the child, and the techniques developed can be utilized and expanded for the analysis and developmental study of father-infant and infant-infant affection.

Since we can measure neonatal and infant affectional responses to mother surrogates, and since we know they are strong and persisting, we are in a position to assess the effects of feeding and contactual schedules; consistency and inconsistency in the mother surrogates; and early, intermediate, and late maternal deprivation. Again, we have here a family of problems of fundamental interest and theoretical importance.

If the researches completed and proposed make a contribution, I shall be grateful; but I have also given full thought to possible practical applications. The socioeconomic demands of the present and the threatened socioeconomic demands of the future have led the American woman to displace, or threaten to displace, the American man in science and industry. If this process continues, the problem of proper child-rearing practices faces us with startling clarity. It is cheering in view of this trend to realize that the American male is physically endowed with all the really essential equipment to compete with the American female on equal terms in one essential activity: the rearing of infants. We now know that women in the working classes are not needed in the home because of their primary mammalian capabilities; and it is possible that in the foreseeable future, neonatal nursing will not be regarded as a necessity, but as a luxury – to use Veblen's term – a form of conspicuous consumption limited perhaps to the upper classes. But whatever course history may take, it is comforting to know that we are now in contact with the nature of love.

First published in American Psychologist, 1958, 13, 673–685.

Chapter 3
The Japanese and Japanese Monkeys: Dissonant Neighbors Seeking Accommodation in a Shared Habitat

Yoshihisa Mito and David S. Sprague

3.1 Introduction

The ancestors of both the Japanese monkey (*Macaca fuscata*) and humans (*Homo sapiens*) arrived in the Japanese archipelago during the Pleistocene epoch, probably via the Korean Peninsula when a land bridge connected Japan to the Asian mainland. According to fossil evidence, the earliest recorded ancestors of the macaque were in the Japanese archipelago nearly 430,000 years ago (Aimi 2002). The oldest subfossil of Japanese people has been found in a geologic stratum approximately 35,000 years old. Thus, humans and monkeys have lived together in Japan for more than 35,000 years. The purpose of this chapter is to sketch out the historical transformations in this long and fraught relationship between the people and monkeys of Japan. It is not the intent to fully review the already extensive literature on human-monkey relations in Japan, written in both Japanese (e.g., Hirose 1991; Mito 1992, 2011; Mito and Watanabe 1999) and English (e.g., Knight 1999, 2003, 2005; Ohnuki-Tierney 1987). Rather, the intent is to attempt to identify a few crucial values that we believe have characterized the intimate yet conflicted attitude of the Japanese people toward a fellow primate species with which they share a habitat.

Japan provides an example of a global issue where humans and nonhumans have come to an accommodation over their mutual existence within their overlapping habitats. We present this chapter as a case study of how a culture and an animal have interacted with each other. We do not believe that the Japanese case can be a model for any other culture. Nor do we claim that the Japanese have some sort of special

Y. Mito
Formerly Japan Monkey Centre, Inuyama, Japan

D.S. Sprague (✉)
Ecosystem Informatics Division, National Institute for Agro-Environmental Sciences, Tsukuba, Ibaraki 305-8604, Japan
e-mail: sprague@niaes.affrc.go.jp

S. Radhakrishna et al. (eds.), *The Macaque Connection: Cooperation and Conflict between Humans and Macaques*, Developments in Primatology: Progress and Prospects 43,
DOI 10.1007/978-1-4614-3967-7_3,

rapport with nonhuman animals or that the Japanese case is any more or less unique than the distinctiveness found in any culture. Rather, we wish both Japanese and non-Japanese readers to reflect on the deep contradictions in the relationship between humans and nonhuman primates and to reach a fuller understanding of how each of our cultures reaches an accommodation with nonhuman animals both ecologically and symbolically.

From the Neolithic, monkeys were hunted as food or venerated for their spiritual powers. As history progressed, however, the Japanese view of monkeys changed, along with the psychological and spiritual culture of the Japanese people. Once treated as godlike, people began to think of monkeys as equals, or less, as objects of ridicule or even as a scapegoat (Ohnuki-Tierney 1987). In daily life, monkeys continued to be hunted as a natural resource or despised as pests harming crops as people settled in villages to farm. Today, monkeys have become subjects of scientific research into their natural ecology as well as for the management of their populations. Nevertheless, most Japanese people continue to be keen observers of the macaque. They still retain their faith in the special powers that these monkeys are believed to possess, and the species has thus reserved its special symbolic place in Japanese culture.

3.2 Prehistoric to Early Historic Eras

As the ice age receded, the people living in Japan at the time began to develop very elaborate cultures (Totman 2000). The Neolithic Jomon Era (circa 16,000–2,400 BP) is named for the Jomon style of earthenware, some of the earliest pottery in the world. The Jomon Era saw rapid expansions of many settlements throughout Japan that were very large for a mostly hunting and gathering society. However, the ecological impact of humans intensified with the advent of agriculture and metal tools. Rice cultivation arrived in Japan at about 2,400 BCE to reinforce the trend toward the development of a more agriculturally based society. By the third century CE, early states grew strong enough to build giant funerary, tumulus monuments, known as kofun. Later, under the influence of continental cultures from Korea and China, the aristocracy of Japan developed a highly refined court culture in palatial cities, culminating in the building of the capital city at Kyoto in 794 CE where court culture flourished.

Archeological evidence shows that humans and monkeys had always lived in close proximity. The Jomon settlements left a large number of shell mounds throughout Japan, from which archeologists have unearthed large amounts of animal remains (Habu 2004). The presence of monkey bones in the shell mounds, along with those of many other species of mammals, suggests that monkeys may have been hunted as food by the Jomon people (Fig. 3.1; Hongo et al. 2002). The Jomon people also carved monkey bones to produce decorative ornaments. Archeologists have found what appear to be earrings made from vermillion painted monkey radii, colored red with a hole drilled in it from the Yoshigo kaizuka site in Aichi Prefecture.

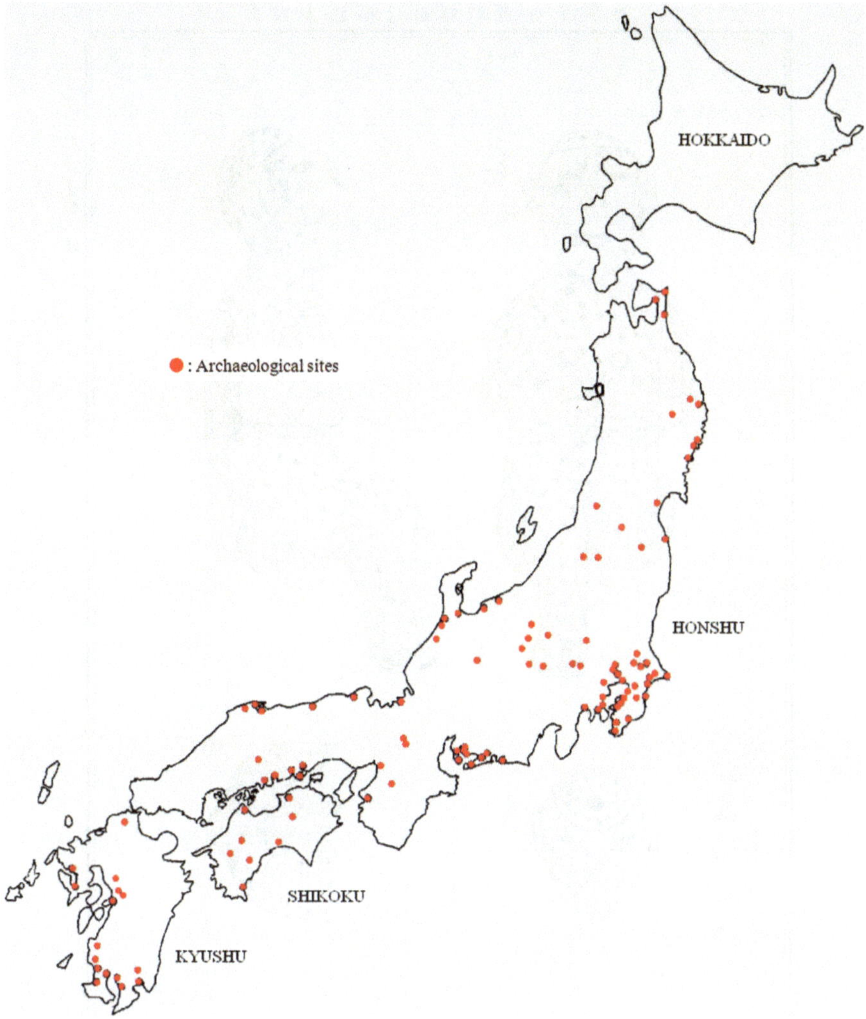

Fig. 3.1 Shell midden sites of the Neolithic Jomon period containing remains of monkeys (Modified from Hongo et al. (2002))

A clay figurine of a female monkey from a Late Jomon site (circa 3,500 BP) of Aomori Prefecture shows that humans had closely observed monkeys (Fig. 3.2). The figurine has hands held over the stomach, nipples, ischial callosities, anus, female genitalia, and the short-tail characteristic of Japanese macaques. The figurine has holes drilled on either side and may have been worn as an amulet for seeking bountiful harvests or easy birth.

The human familiarity with monkeys continues in later periods. Monkeys have been found associated with the kofun tumulus monuments. The kofun were often

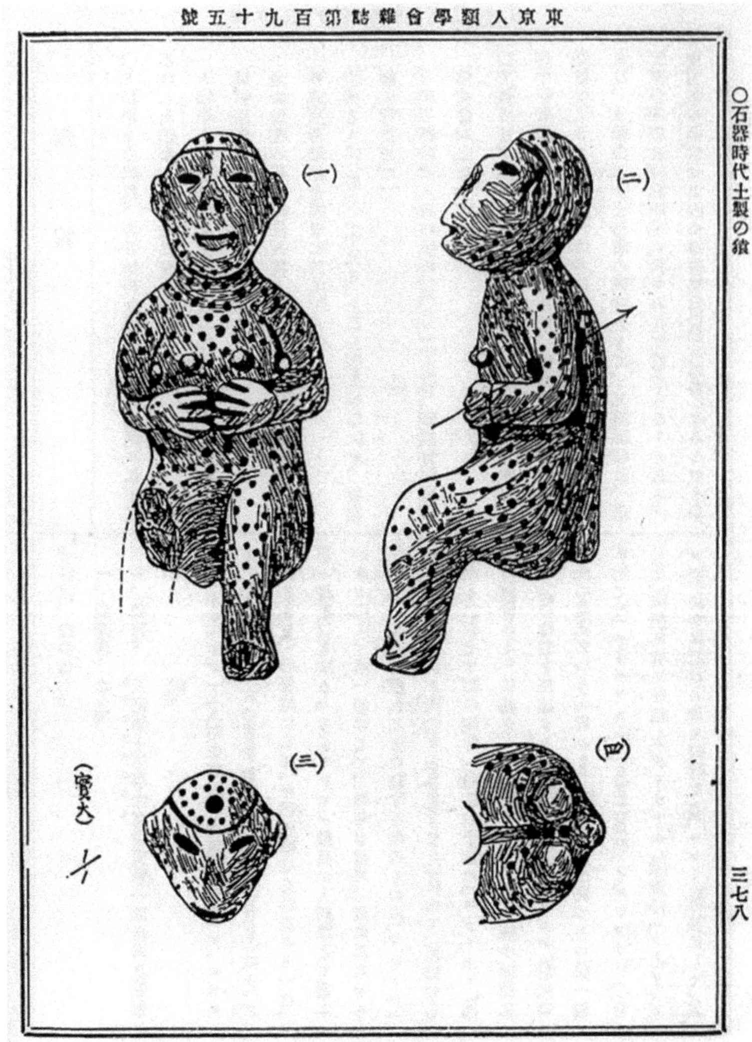

Fig. 3.2 Figure of a ceramic figurine of a female Japanese monkey excavated from a late Jomon site, in a report published in the Tokyo Journal of Anthropology no. 195 (1902) (Reproduced with the permission of the Anthropological Society of Nippon)

accompanied by large numbers of terracotta figurines, called haniwa, depicting soldiers, horses, houses, or other items that may have been important to the dignitaries entombed in the tumulus. A monkey haniwa has been discovered from the Dainichizuka kofun in Ibaraki Prefecture, dated to approximately 550 years CE. The monkeys are depicted accurately with faces painted red (Fig. 3.3).

Fig. 3.3 A haniwa figurine, about 15 cm high, unearthed from a kofun tumulus monument dated approximately 550 CE, the Dainichizuka kofun in Ibaraki Prefecture. Detail from a photo of the Tokyo National Museum: TNM Image Archives (Reproduced with the permission of the Tokyo National Museum)

The arrival of Buddhism to Japan altered radically the human view of animals. Buddhism encourages respect for all life and discourages people from hunting or eating animals. This led to official proscriptions against hunting and eating animals. In a decree dated the fourth month of the fourth year of the Tenmu Era (676 CE), the emperor ordered that the following animals must not be eaten: cattle, horse, dog, monkey, and chicken (Ohnuki-Tierney 1987). Why monkeys were included in this list is somewhat mysterious. Cattle, horses, dogs, and chickens are domesticated animals. Did humans also keep monkeys? This may have meant that monkeys were kept as pets. Alternatively, the emperor may have included monkeys in the decree because monkeys were similar to humans.

The flowering of aristocratic culture nurtured the arts, including painting. Some of the earliest depictions of monkeys seem to have been products of practicing artists. Artists sometimes practiced drawing on fragments of broken pottery because paper was too valuable for practicing drawing. Four monkeys and many characters are drawn with charcoal on the back of a 20-cm plate, excavated from an estate of the emperor's grandson, the Nagayaou in Nara Prefecture, dated to the eighth century CE (Tatami 1988). The artist undoubtedly had the opportunity to closely observe monkeys, for they are drawn animatedly and accurately, even showing the large cheek pouches (Fig. 3.4).

Fig. 3.4 Detail from a drawing of a monkey on a plate excavated from the Nagayaou estate site in Nara, one of Japan's ancient capitals. The plate is about 20 cm in diameter. The drawing shows the monkey's cheek pouch. Photo from Tatsumi (1988) (Reproduced with the permission of the Japan Monkey Centre)

3.3 Middle Ages: From Twelfth to Sixteenth Century

The beginning of the Japanese Middle Ages was marked by the decline of the aristocracy and the rise of the samurai, one of which became a shogun, the military ruler of Japan who presided in parallel to the greatly weakened emperors (Totman 2000). Old values were discarded, and the shoguns came to rule a feudal regime in which they held formal authority over everything in the mountains, rivers, and all of nature. However, regional lords ignored the shoguns and indulged in endless fighting during the Japanese era of warring states. Despite political turmoil, the arts continued to flourish in many ways, and artists continued to depict the changing relationships between humans and animals, including monkeys, in their own creative ways.

Humans probably kept monkeys and may have walked monkeys on leashes from ancient times. A highly developed form of trained monkey performance, the "saru-mawashi," appeared in the twelfth century. A scroll painting from the fifteenth century, the "Sanjuniban Shokunin Itaawase Emaki," depicts a saru-mawashi street performer. This scroll depicts 32 practitioners of various crafts and professions, and a saru-mawashi is among these professions (Fig. 3.5). The word saru-mawashi refers to both the human trainers and the performance of the monkeys. The saru-mawashi removed monkeys from their original habitat and anthropomorphized their actions to make them play a role within human culture. The strictly trained monkey performed tricks and skits parodying human actions. The saru-mawashi still exists today as a form of traditional art (Ohnuki-Tierney 1987).

Fig. 3.5 Detail from the scroll painting titled Illustrations of Thirty-Two Professions (Sanjuniban Shokunin Itaawase Emaki) from the Muromachi period, fifteenth century, of the Japanese Middle Ages. The man is labeled a "Saru Hiki," a monkey leader or puller (Reproduced with the permission of the Tenri Central Library Tenri University)

Anthropomorphized monkeys appear in other art forms as a metaphor or as vehicles of satire. Art historians are familiar with a set of scroll paintings titled Frolicking Animals and Humans or "Chouju Jinbutsu Giga" (Fig. 3.6). In a set of four scrolls produced between the twelfth and thirteenth centuries, monkeys, rabbits, and frogs are shown engaged in human activities, playing games, bathing, or carrying out religious ceremonies. The drawing style is that of Buddhist artistry, but the scenes are comical and appear to satirize the Buddhist priests. Monkeys are depicted wearing clothes and giving instructions to rabbits and frogs in some scenes, but in another, the monkeys in priestly garb worship a frog deity. Curiously, not all animals are anthropomorphized in these paintings. Deer and wild boar appear as wild animals under the control of the anthropomorphized animals.

Fig. 3.6 Details from the scroll painting titled Frolicking Animals and Humans (Chouju Jinbutsu Giga). Produced between the twelfth and fifteenth centuries. Attributed to the priest Tobashoujou Kakuyuu and others (Reproduced with the permission of the Toganosan Kozanji Temple)

Another scroll painting from the sixteenth century actually depicts a major change in the relationship between humans and monkeys. Titled the "Fujibukuro Soshi," or The Story of the Wisteria Basket, and drawn by a female artist, Tosa Mitsuhisa, the painting tells the tale of a young human woman marrying a monkey (Fig. 3.7). However, this marriage is not a happy one. The story starts with an elderly man, tired by arduous fieldwork, declaring that he will allow anyone who would do the fieldwork for him to marry his daughter. Monkeys appear from the mountains to do the fieldwork, and the elderly man allows a monkey to marry his daughter. She is taken to live with the monkeys in the mountains, where she is placed in a basket made of wisteria, hence the name of the story. The monkeys rejoice, but the young lady is saddened. Then, hunters come to rescue the young lady and kill all the monkeys. Returned to the human realm, the young lady marries one of the hunters.

This story can be interpreted as portraying the process through which humans divorced themselves from monkeys. People had become aware that the age of mystical participation had ended. The artist explains the difference between humans and monkeys and also uses monkeys as a metaphor to critique upstart humans.

Fig. 3.7 Detail from the scroll painting titled Story of the Wisteria Basket (Fujibukuro Soshi), produced about 1500 in the Muromachi period of the Japanese Middle Ages, 1,360.7 cm long and 19.8 cm wide. Attributed to Tosa Mitsuhisa, a female artist (Reproduced with the permission of the Suntory Museum of Art)

Fig. 3.8 Examples of the stable monkey custom, photographed by Y. Mito. The hands of monkeys are attached to a post of a stable (*left*), and a monkey cranium is placed inside a stable

Another new role for monkeys is found in one of the panels of this scroll painting, in which a monkey captured alive is shown leashed in a stable with a horse. The painting may have depicted the origin of the stable monkey custom. The stable monkey custom, Umaya-zaru, began when samurai started to keep pet monkeys in stables. The stable monkey was believed to take on the diseases and misfortunes of horses. This custom has lasted for 500 years; the practice of placing monkey parts in stables continues in some parts of Japan today (Fig. 3.8).

3.4 Early Modern Period: Seventeenth Century–1868

The early modern period of Japan is defined to have started with the final victory of one of the great lords, Tokugawa Ieyasu, who succeeded in unifying the country under his rule (Totman 2000). He received the title of shogun in 1603 and established the political capital at Edo, the city that later became Tokyo. Thus this period in Japanese history is often called the Tokugawa Era or the Edo Era. Still fundamentally a feudal political system, 15 generations of Tokugawa shoguns ruled until 1868 by controlling the provincial lords who continued to have the power of life and death over their domains.

The Tokugawa shogunate enforced a long period of peace and international isolation during which Japan developed rapidly both socially and economically. The population and commerce increased greatly, placing more pressure on Japan's agricultural capacity and natural resources. The Japanese relationship to nature, while still retaining many religious aspects, became more objective and pragmatic.

The greatest artist to draw monkeys worked in the late Edo Era. Mori Sosen (1747–1821) made a reputation for drawing monkeys realistically (Kawano 1972; Kimura 1975). Schooled in painting in the Chinese style, Mori was an artist who lived in a city but specialized in painting animals. To learn to draw monkeys, Mori purchased a monkey from a hunter and kept it in his house to serve as his model. However, an acquaintance told him that the monkeys he drew did not appear natural. To learn to draw monkeys realistically, he traveled to observe monkeys in their natural environment. He lived in the mountains to observe monkeys for 3 years and learned to depict monkeys as they actually lived. By becoming a monkey himself, Mori Sosen drew monkeys with an intuitive understanding that is unparalleled even today. Those who saw his drawings declared his ability to be godlike.

Even for a modern primatologist, Mori's paintings are uncanny in expressing the essence of macaque posture and behavior (Fig. 3.9). In recognition of his observational

Fig. 3.9 Illustrations of monkeys by the late Edo Era artist Mori Sosen. Two volumes of sketches of monkeys in charcoal survive. The two *left figures* are copies drawn in the Meiji Era, reproduced with the permission of the Japan Monkey Centre. The figure on the *right* is a painting by Mori Sosen that is used on the cover of Primate Research, the Japanese language journal of the Primate Society of Japan, reproduced with the permission of the Primate Society of Japan

ability and artistry, a drawing by Mori Sosen was chosen as the cover picture from the first (December 1985) issue of *Primate Research*, the Japanese language journal of the Primate Society of Japan. Given his keen observations of monkeys, carried out during long years spent with them and which formed the basis of his endearing studies, we would like to consider Mori Sosen the world's first primatologist.

A serious practical problem with monkeys was the conflict between farmers and wildlife. Wild animals damaged crops, inflicting serious damage on the livelihood of farm communities. There are even accounts of a famine that was reportedly caused by wild boars (Kikuchi 2000, 2009). In 1749, 3,000 people died in Aomori Prefecture in the "wild boar famine" when boars destroyed many farms. While the ecological causes are undoubtedly complex, some historians believe that a misguided wildlife management program of the local lord contributed to the problem. The lord had encouraged the hunting of wolves to protect his horses, resulting in an increase in the number of the prey animals, including wild boar, deer, and monkeys.

Historical records show that farmers had limited means to protect their fields from wildlife (Hosoi 1987; Oka 1979; Tsuchiya 1988). They were prohibited from hunting and ate meat only rarely. They were not allowed to own firearms or other weapons and could only chase away wild animals. Farmers deployed many defensive mechanisms to guard their fields: fences and look-out huts around the perimeter, noise makers, dogs, and feces of animals, especially of wolves. The shishi-odoshi, a bamboo sound maker seen in traditional gardens, was originally used for frightening away animals (Mito 1989). Hunting of wild animals was carried out by professional hunter communities that had permission to hunt as well as use firearms or other weapons (Chiba 1969, 1992). The most famous of these hunting communities were the Matagi of northern Japan (Oota 1979). The Matagi observed a strict hunting ethic, mandated by observances of gratitude and fearful awe to the mountain god for the bounty of nature allowed them. The Matagi mainly targeted bears. Bear gallbladder and skins were the taxes paid by hunters to the lords of Morioka and Akita provinces. The Matagi were allowed to consume some of their catch but strictly prohibited from trading on their own because the lords controlled the trade in valuable animal products (Muto 1969). Monkeys were hunted sometimes because monkey gallbladder was a medicine highly valued by farm communities.

3.5 Modern Era: 1868–1945

Feudalism ended with the dissolution of the Tokugawa shogunate in 1868 and the beginning of the modernizing government under Emperor Meiji (Totman 2000). Industrialization commenced, and Japan set forth on the road to a capitalist, market economy. The liberalization and commercialization of Japanese society following the end of feudalism led to a vast expansion in the trade in animal and forest products. The mode of taxation shifted from kind to money. This had the effect of accelerating the monetization of natural resource usage and undermined traditional controls on land use and resource harvesting.

Fig. 3.10 Photograph taken after a monkey hunt in Tottori Prefecture in 1923. The man in the center is holding a rifle (Reproduced with the permission of the Japan Monkey Centre)

Forestry increased in importance all over Japan. The expanding economy and population created greater need for wood and fuel, much of which continued to be produced domestically in Japan. In particular, until fossil fuels attained common usage following the Second World War, much of household fuel continued to be charcoal and firewood (Higuchi 1960). The expansion of forestry meant greater turnover in the habitats for forest-dwelling animals.

In addition, with the end of feudal occupational strictures, hunting was permitted for all citizens (Chiba 1969, 1992). Hunting accelerated greatly as less expensive Japanese-made rifles replaced muzzle-loading matchlock guns. Wild animal populations decreased dramatically, and the Japanese wolf became extinct. Wild animal products became valuable as food, clothing, and medicine. There was high demand for these products, and they were sold at high prices. People believed that nature was bountiful, and "insects, fishes, birds, beasts" were plentiful. In addition, the government did very little to preserve nature or wildlife. These factors contributed to the regional extinction of many species. The hunting of monkeys also accelerated. Not only were monkeys pest animals, their meat, fur, and gallbladder could be sold at high prices. Furthermore, a medicine made from monkeys – black-roasted monkey cranium ground into a powder – was popular in villages as a cure for maladies connected to the head and women's ailments. From the Meiji, Taisho, and into the Showa periods, monkeys were hunted throughout Japan (Fig. 3.10). A Matagi

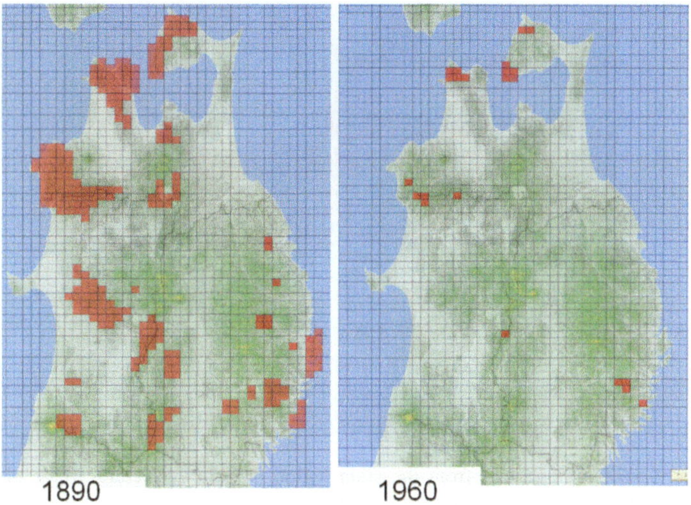

Fig. 3.11 Distribution of monkeys in the three prefectures of Aomori, Akita, and Iwate in the northern Tohoku region of Honshu Island in Japan (Figure by Y. Mito)

once shot 32 monkeys in a single hunting expedition by taking advantage of the tendency of monkeys to travel in groups (Oota 1979). Figure 3.11 shows distribution maps of monkey populations in the Tohoku region where their numbers greatly diminished during this period.

3.6 Present Period: 1945–

Recovering from the devastation of the Second World War, Japan rapidly developed its industries, and environmental pollution worsened. Fortunately, environmental protection was reestablished, and the values of nature protection have taken deeper root in Japanese society. The period immediately following the end of the Second World War appears to be the phase when Japanese people had least contact with monkeys. Macaque populations had greatly reduced over much of Japan, and in 1947, monkeys were finally given legal protection and placed on the list of non-hunted animals under the laws regulating hunting. But even before the hunting ban, according to government statistics, fewer monkeys were hunted during the immediate postwar years. However, serious crop damage by monkeys started again in the 1970s. The hunting of monkeys, not as a resource or for sport but as a pest, increased dramatically, exceeding 10,000 animals per year in 1988. Primatologists have found it difficult to fully explain why crop damage by macaques has seen such a dramatic resurgence. We believe it is the result of a combination of various factors – rural depopulation due to younger people congregating in the cities, the aging rural population, massively expanded conifer plantation forestry, and the loss of traditional rural land uses resulting in closer proximity between farms and forests (Muroyama

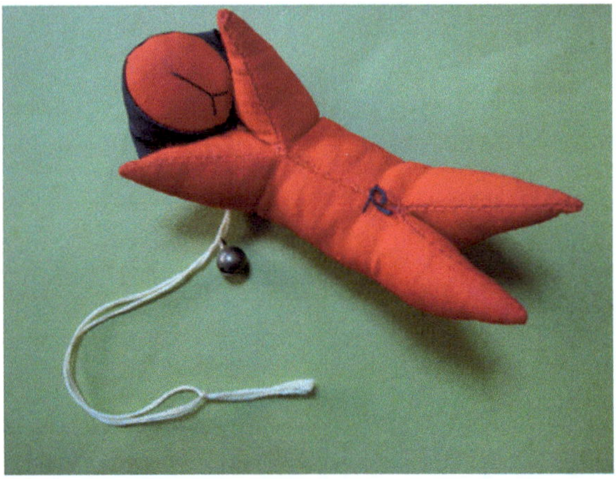

Fig. 3.12 "Sarubobo" figurines from the Hida region of Gifu Prefecture. Made of cloth, the figurines are protective amulets in the form of monkeys, now often sold as souvenirs to tourists, even at the Nagoya Airport (Photo by Y. Mito)

and Yamada 2010; Sprague 2002; Sprague and Iwasaki 2006, 2010; Suzuki and Muroyama 2010). It is unfortunate that the interface between humans and monkeys in much of Japan is once again characterized by conflict.

Nevertheless, monkeys continue to play large roles in cultural practices in many parts of Japan. Monkeys are central characters in many traditional children's stories. In the Monkey Jizo legend of Yamagata Prefecture, an old farmer takes a nap during fieldwork. A group of monkeys mistake him for a Buddhist Jizo statue and bring him many rich offerings. The old man awakes to find himself enriched. Monkey figures still serve as modern protective amulets. For example, the "sarubobo" of Gifu Prefecture is an infant monkey figurine that serves as a protective amulet to avoid misfortunes (Fig. 3.12). It is now a popular gift item for tourists.

A new institution created in the postwar period, specifically intended to bring monkeys and people together, were monkey parks. Monkeys at that time were elusive and unusual animals that had to be enticed with food offerings to visit the parks. Monkey parks were set up in many parts of Japan for various reasons, such as tourism, to move monkeys away from human habitations and to help in the recovery of monkey populations (Mito 1995). Unlike zoos, tourists can visit a monkey park to mingle among free-ranging monkeys and often directly interact with monkeys by feeding them (Knight 2005). Several parks were set up with the cooperation of primatologists who carried out some of the most important early research on the social behavior of monkeys in these parks, in particular, Takasakiyama in Oita Prefecture and Arashiyama in Kyoto.

Japanese primatologists took full advantage of living in a nation with monkeys (Yamagiwa 2010a). Many generations of Japanese students started their research career by being sent off by their professors to nearby mountains to watch monkeys. Partly because of the excellent observational opportunities at monkey parks, scientific research on Japanese monkeys quickly achieved world-renowned results and focused attention on many fundamental aspects of macaque behavior such as the matrilineal dominance rank system and behavioral variability and cultural traits like potato washing. Later, just as Mori Sosen wanted to draw naturalistic monkeys, a new generation of primatologists sought out field sites to study Japanese monkeys in entirely natural habitats, further enriching the body of scientific literature on Japanese macaques (Yamagiwa 2010b).

The scientific study of animals in Japan does not exclude Japanese customs. One example of the confluence of tradition and science is the Buddhist Kuyo memorial rite carried out in many research institutes in Japan for the animals studied at the institute (Asquith 1986a). The ceremony is often named after the subject being memorialized, and "Saru Kuyo" ceremonies are carried out at the Japan Monkey Centre and the Kyoto University Primate Research Institute. In Buddhist theology, the rite is for comforting the departed soul, sending them to the next world, and wishing them a good life in their next life. Through the ceremony, humans apologize for taking the life of animals and express gratitude for the benefits gained by humans.

3.7 Conclusion: The Japanese Monkey, a Neighbor to the Japanese People

The Japanese people have found multiple and often contradictory meanings in the presence of monkeys in Japan. These multiple meanings persist today and interact with each other in complex contexts. The spiritual power of monkeys may have diminished in Japanese culture, but the stable monkey custom remained until recently in many parts of Japan. Not only research institutes but many zoos also carry out annual Kuyo ceremonies to express gratitude and gain the indulgence of animal spirits. Government policies to manage and control animal populations are tempered by heightened awareness for nature protection. Even though thousands of monkeys are culled as pests every year, a general antipathy to killing animals persists, and in some parts of Japan, primatologists have heard of the fear of some hunters of a curse from killing monkeys because of the similarity between humans and monkeys.

It is tempting to seek in religious beliefs the foundation to Japanese attitudes toward nature. However, no simple, one-to-one relation exists between Japanese religious beliefs and particular attitudes about monkeys in Japan. We defer to scholars more knowledgeable about religion for a detailed analysis but suggest a few basic values that the majority of our Japanese readers may agree continue to exist widely among the Japanese people.

The Japanese people consider that all animate beings possess a spirit and a personality. The Shinto religion, worshipped in Japan from ancient times, has brought the animistic view of nature into present-day Japanese culture, in which the forces of life and spirit are seen within any and all things in the natural world, not only in living things, animals and plants, but also in nonliving stones and earth and water. When the philosophical religion of Buddhism arrived in Japan via the Chinese continent, some Japanese Buddhists introduced into Buddhism the philosophy of original enlightenment (*hongaku shiso*) where the mountains and rivers and all living creatures possessed the Buddha nature (*bussho*) (Sueki 1997). Stones, earth, mountains, rivers, lightening, typhoon, the sun and moon, grasses and trees, insects, birds, and beasts, and of course, humans were all placed on the same horizon and considered to possess a soul and were simultaneously god (*kami*) and Buddha spirit (*hotoke*).

Immersed within this culture, the Japanese people freely anthropomorphize and insert emotional significance into many things. The historical paintings and crafts introduced above are a part of the evidence for this attitude of the Japanese. This is not only done with animals and plants but even with tools. Every tool is recognized to have a unique personality, and the fundamental ethic of Japanese craftsmanship lies therein. Japanese people are moved by the fate of every object, imbued with a spirit, fulfilling its destiny given to it by heaven. In a recent example, Japanese people were moved, inspired by, and even shed tears for the space probe "Hayabusa" that returned from a 7-year long journey to an asteroid, and fulfilled its mission by burning to extinction as it entered the Earth's atmosphere to successfully deliver its tiny payload. Kuyo rites recognize the significance and sacrifice of many things. There are sewing-needle Kuyo and doll Kuyo ceremonies, to name only a few. Kuyo rites can be merely the seeking of absolution, but it also admonishes people not to treat things casually or callously. That is why research institutes perform Kuyo memorial rites for monkeys.

Anthropomorphism has been critiqued by many philosophers (Asquith 2011), and its role in interpreting nonhuman animals has been considered highly problematical, although some researchers have defended the role of anthropomorphism in the study of nonhuman animal behavior (Fisher 1996; Keeley 2004). While Western researchers also exhibit anthropomorphism (Rees 2007), it has been claimed that Japanese primatologists practice anthropomorphism more self-consciously, as compared to their Western counterparts (Asquith 1986b). We wish to emphasize here that a critique of Japanese anthropomorphism should perhaps take into account its philosophical basis and its sociocultural roots. The anthropomorphism evident in the human-monkey interrelationships, as seen in Japan, is based on an assumption of a fundamental equivalence among living things that allows humans to assimilate to some degree with monkeys without precluding the opportunity for dispassionate observations of the other. This dualism may be seen in the scroll painting of the Frolicking Humans and Animals where animals are anthropomorphically assimilated with humans and yet objectified in a highly detailed pictorial medium. The naturalistic paintings by Mori Sosen are simultaneously intimate and anatomically accurate.

This dualism also may be related to why early Japanese primatologists sought out opportunities to observe primates as closely as possible. Within the context of a fundamental equivalence between human and nonhuman social animals, the observer was expected to assimilate to some degree with the subject while proximity permitted accurate observations. Early primatologists achieved proximity by feeding monkeys at provisioning sites. Later, primatologists spent many years habituating monkeys without provisioning and establishing more naturalistic study sites, patiently waiting for the time when a field researcher could sit next to a wild monkey on the same ground level. Positioned closely to monkeys, the primatologists gave individual monkeys the names of Greek gods because they quickly recognized the personalities of the monkeys with whom they interacted. It is not surprising that postwar Japanese primatologists readily recognized that monkeys possessed a complex social system.

To be sure, a great deal of casual anthropomorphism occurs in Japan. Knight (2005), in his study of monkey parks, characterizes them as sites of pervasive anthropomorphism where visitors are supposed to be able to encounter the grateful monkey of Japanese folklore in the act of hand-feeding the monkeys. However, Knight (2005) points out how visitors quickly discover that the monkeys ignore human etiquette. Visitors are forced to study monkey social behaviors to feed a favorite individual monkey by devising a handout strategy counteracting the monkey troop's dominance hierarchy. Intimate interaction negates casual anthropomorphism, forcing visitor and monkey into keenly observing each other, in what Knight (2005) calls the meeting of dissonant socialities.

All of these multifaceted meanings gathered into one culture may only seem confusing, and it is hard to advocate a cultural basis for the Japanese people to continue their coexistence with monkeys. We confine ourselves to suggesting that Japanese culture retains the concept of coexistence with monkeys based on a sense of equivalence among living things in this transient world. This recognition of nature is part of Japanese tradition and has fostered many aspects of Japanese culture and society, including primatology.

While nature conservation policy may be mainly the work of scientists and the government, it is important that Japanese people continue their relationship with monkeys as neighbors with their own personalities, based on a culture of "anthropomorphized personalities." This neighbor is a difficult one, and maintaining a working relationship requires continuous and endless effort. Some of the community conservation efforts being developed around Japan may represent the hard work necessary to reestablish the human-monkey relation in Japan (Suzuki and Muroyama 2010). These community conservation programs call for the fine-grained planning of many detailed practices needed to manage village environments based on a better understanding of how both humans and monkeys see the environment that they share, such as by reducing crop residues left in fields or training "monkey-dogs" to guard villages (Izawa 2005). Maintaining a working relationship with wildlife is a time-consuming and difficult effort, but it is one way to achieve coexistence with mutual benefits, *kyozon-kyoei* in Japanese, for humans and monkeys.

For non-Japanese readers, we submit that to look deep into one's culture is to gain an awareness of the history and values that underlie the relationship between humans and wildlife. Whatever the relationship, purely technocratic values are unlikely to gain public support for either the protection or management of primate populations. We hope the Japanese case provides some insights, but the actual solutions to wildlife issues need to be formulated within the cultural context of each place where humans and primates coexist.

Acknowledgments This chapter was originally prepared by Y. Mito and D. S. Sprague for the symposium titled "Japanese Monkeys and the Japanese People: History and Issues in Japanese Monkey Conservation," held at the XXIII Congress of the International Primatological Society, in

Kyoto, Japan, in September 2010. We thank the members of the Conservation Committee of the Primate Society of Japan, especially Dr. T. Oi, for organizing the symposium. Finally, we thank Dr. Oi and an anonymous reviewer for their many useful comments of an earlier draft.

References

Aimi M (2002) The oldest Japanese monkey. Asian Paleoprimatol 2:13–19

Asquith PJ (1986a) The monkey memorial service of Japanese primatologists. In: Sugiyama Lebra T, Lebra WP (eds) Japanese culture and behaviour: selected readings. University of Hawai'i Press, Honolulu, pp 29–32

Asquith PJ (1986b) Anthropomorphism and the Japanese and Western traditions in primatology. In: Else JG, Lee PC (eds) Primate ontogeny, cognition, and social behavior. Cambridge University Press, Cambridge, pp 61–71

Asquith PJ (2011) Of bonds and boundaries: what is the modern role of anthropomorphism in primatological studies? Am J Primatol 73:238–244

Chiba T (1969) Hunting lore research [Shuryou-denshou-kenkyu]. Kazama Shobo, Tokyo (in Japanese)

Chiba T (1992) Hunters of Japan and their activities [Nihon no shuryousha to sono kodo]. In: Koyama S (ed) Hunting and fishing [Shuryo to Gyoro]. Shuzankaku Shuppan, Tokyo, pp 240–254 (in Japanese)

Fisher JAF (1996) The myth of anthropomorphism. In: Bekoff M, Jamieson D (eds) Readings in animal cognition. MIT Press, Boston, pp 3–16

Habu J (2004) Ancient Jomon of Japan. Cambridge University Press, Cambridge

Higuchi K (1960) History of Japanese charcoal use [Nihon Mokutan-shi]. Zenkoku Nenryo Kaikan, Tokyo (in Japanese)

Hirose S (1991) *Monkeys and the Japanese* [Saru to Nihonjin]. Dai-ichiShobo, Tokyo

Hongo K, Fujita M, Matsui A (2002) Time-course changes in the distribution of Japanese monkey, *Macaca fuscata* based on those unearthed on the ground from ancient remains. Asian Paleoprimatol 2:1–12

Hosoi K (ed) (1987) Morioka-han (17th century), Miscellaneous records of the Morioka Domain [Morioka-han Zassho] no. 1~4. Morioka-shi Shakaikyoiku-iinkai, Morioka (in Japanese)

Izawa K (2005) Complete manual for monkey measures [Sarutaisaku Kanzen Manyuaru]. Dobutsu-sha, Tokyo

Kawano M (1972) Mori Sosen research preface [Mori Sosen Kenkyu Jyosetsu]. Kokka (950):5–25 (in Japanese)

Keeley BL (2004) Anthropomorphism, primatomorphism, mammalomorphism: understanding cross-species comparisons. Biol Philos 19:521–540

Kikuchi I (2000) Famine [Kikin]. Shueisha Bunko, Tokyo (in Japanese)

Kikuchi I (2009) The wild boar famine and regional society [Inoshishiare to chiikisyakai]. Furusato Nanbu (3):3–19, Nanbu Township Historical Research Society (in Japanese)

Kimura S (1975) Thoughts of Sosen [Sosen Ko]. Kobijyutsu (49):57–70 (in Japanese)

Knight J (1999) Monkeys on the move: the natural symbolism of people-macaque conflict in Japan. J Asian Stud 58:622–647

Knight J (2003) Waiting for Wolves in Japan. Oxford University Press, Oxford

Knight J (2005) Feeding Mr. Monkey: cross-species food 'exchange' in Japanese monkey parks. In: Knight J (ed) Animals in person: cultural perspectives on human-animal intimacy. Berg, Oxford, pp 231–253

Mito Y (1989) In the case of farmers: monkey damage and its counter measures [Nomin no baai – engai to sono taio]. Monkey J Japan Monkey Centre (227/228): 34–39 (in Japanese)

Mito Y (1992) Why is Japanese monkey distribution so limited in the northern Tohoku Region? [Tohoku-chiho hokubu no nihonzaru no bunpu wa naze sukunaika]. Biol Sci [Seibutsu Kagaku] 44:141–158 (in Japanese)

Mito Y (1995) Prospects of the wild monkey parks [Yaen-koen no shochyo to shorai]. Wildl Conserv Japan 1:111–126 (in Japanese)
Mito Y (2011) The livelihoods of people and the changing monkey distribution of Tohoku Japan [Tohoku Chiho no nihonzaru no bunnpu hensen to hitobito no kurashi]. In: Yumoto T, Shirouzu S, Ikeya K (eds) Environmental history of mountains and forests [Yama to Mori no Kankyo-shi]. Bunichi Sogo Shuppan, Tokyo, pp 99–120 (in Japanese)
Mito Y, Watanabe K (1999) The social history of people and monkeys [Hito to Saru no Shakai-shi]. Tokai University Press, Tokyo (in Japanese)
Muroyama Y, Yamada A (2010) Conservation: present status of Japanese macaque population and its habitat. In: Nakagawa N, Nakamichi M, Sugiura H (eds) The Japanese macaques: their plasticity and robustness. Springer, Tokyo, pp 143–164
Muto T (1969) Testimony of the Akita Matagi [Akita Matagi Kikigaki]. Keiyu-sha, Tokyo (in Japanese)
Ohnuki-Tierney E (1987) The monkey as mirror: symbolic transformations in Japanese history and ritual. Princeton University Press, Princeton
Oka M (ed) (1979) A farmer's diary [Hyakusho Denki] written by an anonymous author (17th century). Compil Agric Manuals [Nihon Nosho Zenshu] No. 17. Noson Gyosonn Bunka Kyokai, Tokyo (in Japanese)
Oota Y (1979) Records of the disappearing livelihoods of mountain people: the matagi [Kieyuku Yamabito no Kurashi]. Suiyosha, Tokyo (in Japanese)
Rees A (2007) Reflections on the field: primatology, popular science and the politics of personhood. Soc Stud Sci 37:881–907
Sprague DS (2002) Monkeys in the backyard: encroaching wildlife and rural communities in Japan. In: Fuentes A, Wolfe L (eds) Primates face to face: the conservation implications of human-nonhuman primate interconnections. Cambridge University Press, Cambridge, pp 254–272
Sprague DS, Iwasaki N (2006) Coexistence and exclusion between humans and monkeys in Japan: is either really possible? Ecol Environ Anthropol 2:30–43
Sprague DS, Iwasaki N (2010) Did a habitat bottleneck exist in the recent history of Japanese macaques? In: Nakagawa N, Nakamichi M, Sugiura H (eds) The Japanese macaques: their plasticity and robustness. Springer, Tokyo, pp 345–357
Sueki H (1997) History of Buddhism in Japan [Nihon Bukkyou-shi]. Shincho-sha, Tokyo (in Japanese)
Suzuki K, Muroyama Y (2010) Resolution of human-macaque conflicts: changing from top-down to community-based damage management. In: Nakagawa N, Nakamichi M, Sugiura H (eds) The Japanese macaques: their plasticity and robustness. Springer, Tokyo, pp 359–373
Tatami J (1988) Monkey illustrated plates from the Nagayou estate excavation. Monkey J Jpn Monkey Centre (219/220):4–5 (in Japanese)
Totman C (2000) A history of Japan. Blackwell, Oxford
Tsuchiya T (ed) (1988) The complete agricultural manuals [Nogyo Zenshu] by Miyazaki Yasusada (1697). Iwanami Shoten, Tokyo (in Japanese)
Yamagiwa J (2010a) Research history of Japanese macaques in Japan. In: Nakagawa N, Nakamichi M, Sugiura H (eds) The Japanese macaques: their plasticity and robustness. Springer, Tokyo, pp 3–25
Yamagiwa J (2010b) History and present scope of field studies on *Macaca fuscata yakui* at Yakushima Island, Japan. Int J Primatol 29:49–64

Chapter 4
Songs of Monkeys: Representation of Macaques in Classical Tamil Poetry

Sindhu Radhakrishna

4.1 Introduction

Today, an Internet search for the keywords 'monkeys', 'humans' and 'India' typically yields accounts of growing numbers of rhesus macaques in the small towns and cities of northern India and the threats posed by these monkeys to human life and property. Such reports also describe how people venerate and feed the macaques due to religious sentiments but are increasingly expressing fear and anger over the presence of monkey populations in their neighbourhoods. Although many of these media communications are characteristically sensational, the antagonism against crop-raiding macaques and the cultural tolerance for their lives, which appear to go almost hand in hand, does add up to a rather remarkable scenario. The seeming dichotomy of reverence and revulsion that simultaneously mark human-macaque relations in India (and indeed, in many parts of Asia) has been commented upon by several authors (Knight 1999; Singh and Rao 2004; Saraswat 2010; Radhakrishna and Sinha 2011a, b). This gives rise to an interesting question: Have human-macaque interactions in India always been marked by these deeply polarised elements? Or is this reflective of more proximate causes? Investigating the history of our relations with animals may not always be functional, and yet, it does allow us to appreciate the magnitude of change that has occurred and perhaps learn from some of our old ways. In this chapter, I investigate cultural perspectives of animals as they are revealed through ancient writings; more specifically, I examine a body of classical Tamil literature from southern India to understand how interactions with primates, including macaques, were viewed by people in those ages.

S. Radhakrishna (✉)
National Institute of Advanced Studies, Indian Institute of Science Campus, Bangalore 560 012, India
e-mail: sindhu@nias.iisc.ernet.in

S. Radhakrishna et al. (eds.), *The Macaque Connection: Cooperation and Conflict between Humans and Macaques*, Developments in Primatology: Progress and Prospects 43, DOI 10.1007/978-1-4614-3967-7_4,

4.2 Humans and Animals

Michael Cook, in his stimulating commentary on an Islamic text written in 1326, raises an interesting premise: Monotheistic cultures have little space for animals, whereas polytheism is infinitely more accepting of them (Cook 1999). To support his stand, he compares the case of Islam and Hinduism, specifically with reference to the ways in which monkeys are treated in both religions. Most allusions to monkeys in Islam are contemptuous or, at the best, derisive of the nature of the animal. In contrast, the monkey is a powerful deity in Hinduism, synonymous with loyalty, strength of purpose and shrewdness. Hanuman, the monkey god of Hinduism, continues to be a prominent part of popular culture in many Asian nations, inspiring affection and reverence in equal measures (Wheatley and Harya Putra 1995; Lutgendorf 2007). A similar view is echoed by Ramanujan in his exposition on the different versions of the *Ramayana,* the classical South and Southeast Asian epic poem on the exploits of the great King Rama and his battle with and defeat of the demon King Ravana (Ramanujan 1992). In Valmiki's *Ramayana* (composed between second century BCE and second century CE and usually accepted as the earliest written version of the *Ramayana*), monkeys play a critical role in the ordering of events in the great epic, so much so, that it is difficult to imagine how the story would have proceeded without their presence. Yet, in the Jain rendering of the Ramayana, the *Paumacariya* (circa 2nd–4th CE) by Vimalasuri, monkeys are completely done away with. Instead, their part is played by the *vidyadharas*, celestial beings who are actually related to Ravana. The *vidyadharas* sport monkeys as emblems on their flags; this, the Jain retelling rationalises, is the reason why they were referred to as *vanaras* or monkeys (Ramanujan 1992).

Human societies have always coexisted with animals. The attribution of religious or symbolic significance to certain animals by some societies reflects how the humans therein perceived their place amongst other living species and 'the boundaries between man and beast' (Sterckx 2002). Classical writings about or relating to animals, irrespective of their intended roles as literature, political treatises or scientific discourses, act as rich sources of information about how the premodern world viewed human-animal relations. Such accounts may be from a naturalist viewpoint, zoological descriptions of animals that underline the exotic nature of the species, as seen in Strabo's *Geography* written around 18–24 CE:

> Megasthenes says that the largest tigers are found among the Prasii, even nearly twice as large as lions, and so powerful that a tame one, though being led by four men, seized a mule by the hind leg and by force drew the mule to itself; and that the long-tailed apes are larger than the largest dogs, are white except for their faces, which are black (the contrary is the case elsewhere), that their tails are more than two cubits long, and that they are very tame and not malicious as regards attacks and thefts; and that stones are dug up the colour of frankincense and sweeter than figs or honey; and that in other places there are reptiles two cubits long with membranous wings like bats, and that they too fly by night, discharging drops of urine, or also of sweat, which putrefy the skin of anyone who is not on his guard; and that there are winged scorpions of surpassing size; and that ebony is also produced; and that there are also brave dogs, which do not let go the object bitten till water is poured down

into their nostrils; and that some bite so vehemently that their eyes become distorted and sometimes actually fall out; and that even a lion was held fast by a dog, and also a bull, and that the bull was actually killed, being overpowered through the dog's hold on his nose before he could be released.[1]

Or they may reveal a preoccupation with the utilitarian function of animals, seen, for example, in ancient Chinese texts on animal husbandry, domestication and veterinary treatment (Sterckx 2005). Indian Vedic manuscripts (circa 1500 BCE–400 BCE) are exhaustive in the detail they provide regarding the suitability of certain classes of animals for ritual sacrifices or even for consumption (Smith 1991). References to animals may also be used as symbols or metaphors of 'wild' nature and human dominion over the less civilised aspects of the world. *Gilgamesh*, the Mesopotamian epic (circa 2700 BCE–1800 BCE, widely considered the oldest extant poetry known to the literary world), chronicles the adventures of Gilgamesh, great king of Uruk 'two-thirds divine and one-third human, extraordinary in strength and beauty', and his search for immortality (Kovacs 1985). But at the heart of the poem is the separation between civilised human nature and wild animal nature and the crisis brought about by this rift. The figure of Enkidu, initially more beast than man, later severed from his primal self and finally changed into an oppressor of nature, reflects humanity's need to identify with nature and the repercussions of denying this fundamental link (Barron 2002). The divide between man and animal is more explicit in Judaeo-Christian theology; in his paper, for example, Lynn White (1967) argues that much of early Christian writings emphasise dualism between man and nature and support man's exploitation of nature and animals for his own needs:

And God said, Let us make man in our image, after our likeness: and let them have dominion over the fish of the sea, and over the fowl of the air, and over the cattle, and over all the earth, and over every creeping thing that creepeth upon the earth.[2]

Beast fables, such as the Buddhist *Jatakas*, Aesop's *Fables* and the *Panchatantra*, offer yet another perspective, wherein animal figures are used to convey lessons about human morality and ethics. These stories abound in lordly lions, wily jackals, meddlesome monkeys and foolish crocodiles through whose experiences the author counsels his human readers on the decorum and conduct of life (Pierce 1969). The framing story of the first book of the *Panchatantra* speaks of the amity between a lion and a bull and, through the narration of their broken friendship, inserts some 30 odd tales and sermons on prudent behaviour and the value of honesty and loyalty (Ryder 1925; Olivelle 1997). When the lion king confesses to his counsellor, the jackal, that his henchmen are too afraid to investigate the loud sounds that had frightened him, the jackal replies that they are not to be blamed, as servants take after their masters:

[1] *Geography*: Book XV, Chapter 1, Section 37. Downloaded from http://penelope.uchicago.edu.

[2] Genesis 1:26. The Bible (King James Version).

In case of horse or book or sword,
Of woman, man or lute or word
The use or uselessness depends
On qualities the user lends.[3]

Elsewhere in the book, in the well-known story of *The Monkey and the Crocodile*, when the crocodile's wife demands that he must bring her the heart of the monkey who is his friend, the crocodile tries to refuse saying that the monkey is as dear as an adopted brother. Moreover, the monkey has been gifting them roseapples, so he must not be harmed. To convince his wife, he quotes the proverb:

To give us birth, we need a mother;
For second birth we need another:
And friendship's brother seems by far
More dear than natural brothers are.[4]

The crocodile's wife is not convinced, however, and the rest of the story relates how the crocodile tries to procure the monkey's heart but is finally outwitted by the monkey.

Anthropomorphism is inevitably a large part of these animal representations. Ascribing human qualities to animals is an ancient practice, pervasive amongst all cultures and is, it has been argued, 'built-in to the human repertoire' (Carporael and Heyes 1997). The relation of anthropocentrism to anthropomorphism is a much-debated topic, and several scholars have investigated the subject at length to try and understand why we anthropomorphise (Bacon 1620/1960; Burghardt 1985; Fisher 1991; Kennedy 1992; Mitchell et al. 1997). Closely related to anthropomorphism (and of more interest to this chapter) is the less-discussed concept of zoomorphism. The converse of anthropomorphism, zoomorphism, refers to the assignment of animal characteristics to humans. Zoomorphism plays a significant role in shamanism and totemism; ritual activities allied to these belief systems involve animal relations, communing with animal spirits and transformation into animals (Baldick 2000; Layton 2000; Winkelman 2002, 2004). Indeed, some scholars hold that totemic thought envisages human groups and differences between them, in terms of different categories seen in the animal world (Levi-Strauss 1962, 1963).

4.3 Animal as Metaphor

Human societies have always used animal metaphors to express themselves. From palaeolithical rock paintings through ancient epics to modern novels, animal metaphors have been (and continue to be) used as exemplars to illustrate, substantiate

[3] *The Panchatantra*: Book 1. Translated from the Sanskrit by A W Ryder. Chicago: The University of Chicago Press.

[4] *The Panchatantra*: Book 4. Translated from the Sanskrit by A W Ryder. Chicago: The University of Chicago Press.

and sermonise (Barnett 1990). While anthropomorphism is prominent in such metaphorical usages in classical writings, zoomorphism, although perhaps less common, also occurs alongside. Virgil's *Aeneid* compares the play of Trojan children to dolphins gambolling in the sea:

> so did the sons of Troy their courses weave
> in mimic flights and battles fought for play,
> like dolphins tumbling in the liquid waves,
> along the Afric or Carpathian seas.[5]

and Turnus' anger to a bull that is readying for war (Briggs 1980):

> Such frenzy goads him: his impassioned brow
> is all on flame, the wild eyes flash with fire.
> Thus, bellowing loud before the fearful fray,
> some huge bull proves the fury of his horns,
> pushing against a tree-trunk; his swift thrusts
> would tear the winds in pieces; while his hoofs
> toss up the turf and sand, rehearsing war.[6]

In Hellenistic writings, zoomorphism is more evident in the work of the Epicureans and the Cynics; Lucretius, in *De Rerum Natura*, consistently uses examples of animal behaviour to describe a standard that all humans must aspire to. But even when the position of the Cynic-Epicurean writer is avowedly pro-nature, animal descriptions are notably anthropomorphic: ants are 'fearful of coming age and penury', bees 'house together in one city' and an old horse 'rages idly'[7] (Grant 1969; Gale 2000).

Anthropomorphic and zoomorphic descriptions are relatively more straightforward in classical Indian Sanskrit texts. With its multiplicity of birds and animals that sometimes are dramatis persona, Valmiki's *Ramayana* overflows in anthropomorphic descriptions, but zoomorphism is not far behind. Characters and their behaviours are repeatedly compared to animals, as, for example, Sita is 'fawn-eyed',[8] Narada a 'bull among sages',[9] Rama and Lakshmana 'splendid warriors with the gaze of lions, courageous and strong as lions, majestic, handsome, with the gait of fine bulls, with arms like elephants' trunks',[10] while King Dasharatha caresses his

---

[5] *Aeneid*: 5:592–595. Downloaded from http://www.perseus.tufts.edu.

[6] *Aeneid*: 12:100–106. Downloaded from http://www.perseus.tufts.edu.

[7] *Georgics* (1: 191; 4:155; 3:101) Downloaded from http://www.perseus.tufts.edu·

[8] *Valmiki Ramayana*. Kishkindha kanda: Sarga 1:16. From *The Ramayana of Valmiki: An epic of ancient India, Vol. IV: The Kishkindha kanda* (1994). Translated by Rosalind Lefeber, Edited by Robert Goldman. New Jersey: Princeton University Press.

[9] *Valmiki Ramayana*. Bala kanda: Sarga 1:1. From *The Ramayana of Valmik: An epic of ancient India. Volume1: The Bala kanda* (1984). Translated and Edited by R.P. Goldman. New Jersey: Princeton University Press.

[10] *Valmiki Ramayana*. Kishkindha kanda: Sarga 3: 6–8. From *The Ramayana of Valmiki: An epic of ancient India, Vol. IV: The Kishkindha kanda* (1994). Translated by Rosalind Lefeber, Edited by Robert Goldman. New Jersey: Princeton University Press.

distraught queen as 'affectionately, as a great bull elephant in the wilderness might caress his cow wounded by the poisoned arrow of a hunter lurking in the forest'.[11]

Why is zoomorphism less common than anthropomorphism in premodern literature? Cultural historians have hypothesised that the cultural history of Europe shows three distinct structural layers; the most well-known of these is the historically recent Muslim/Christian layer. This was preceded in prehistorical times by anthropomorphism, which, in turn, succeeded the archaic zoomorphic and totemic layer (Frobenius 1929; Viereck 2002). Each period lasted many thousands of years – Alinei (1997) theorises that anthropomorphism was typical of the Metal Age, while zoomorphism was more representative of the Stone Age. Most certainly, there were transitions and overlaps between the three periods but, inevitably, the cultural influences of the last two periods, at least on language, have been documented better (Viereck 2002). This perhaps may explain why, generally speaking, there are relatively more accounts of anthropomorphism than zoomorphism in premodern writings.

4.4 Classical Tamil Literature and Zoomorphism

Anthropomorphism and zoomorphism display remarkable knowledge of animals and keen interest in their behaviour, but zoomorphism is infinitely more detailed in its observation of 'bestial' qualities that it assigns to human beings (Doniger 2005). One body of classical literature where anthropomorphism is rare and zoomorphism predominates is classical Tamil poetry (Ramanujan 1985; Selby 2011).[12] Tamil is one of the classical languages of India and refers to the Dravidian language that is spoken mainly in the south Indian state of Tamil Nadu. Tamil is also spoken in the nations of Sri Lanka, Malaysia and Fiji. Classical Tamil literature, or *Sangam* literature as it is popularly referred to, comprises Eight Anthologies of Verse (*Ettutokai*), Ten Long Poems (*Pattupattu*) and a grammar composition, the *Tolkappiyam*. Written in Old Tamil (the precursor to modern Tamil) and composed largely between 100 BCE and 250 CE, these texts constitute some of the most original and sensuous poetry found in classical writings (Ramanujan 1967; Zvelebil 1973). The Eight Anthologies are *Narrinai, Kuruntokai, Ainkurunūru, Patirruppattu, Paripāṭal, Kalittokai, Akanāṉūṟu* and *Puṟanānūru. Sangam* poems are classified, on the basis of their themes, as *akam* (interior) and *puram* (exterior) poems. *Akam* poems are love poems, woven around

[11] *Valmiki Ramayana.* Ayodhya kanda: Sarga 10: 4. From *The Ramayana of Valmiki: An Epic of Ancient India. Vol. II: The Ayodhya kanda* (1986). Translated by Sheldon Pollock, Edited by Robert Goldman. New Jersey: Princeton University Press.

[12] Selby's discussion on zoomorphism and the portrayal of animals in Sangam literature in the introductory section of her volume is particularly stimulating, and this chapter owes its genesis to many of the thoughts expressed therein.

the various experiences of love between a man and a woman, while *puram* poems are about action and public life and typically deal with heroic kings, great battles, valorous deeds and the community at large. *Narrinai, Kuruntokai, Ainkurunūru, Kalittokai* and *Akanāṉūṟu* are *akam* anthologies, while *Puṟanānūru, Patirruppattu* and *Paripāṭal* are *puram* anthologies (Ramanujan 1967, 1985; Hart 1999).

Sangam literature is, essentially, landscape poetry. Every poem is ascribed a *tinai* or landscape that designates the mood of the poem. *Akam* poems are represented by five kinds of landscapes – *mullai* (forest), *kurinci* (mountains), *marutam* (pastoral), *neytal* (sandy seashore) and *palai* (wasteland). Each landscape is associated with a particular aspect of love – *mullai* with patient waiting, *kurinci* with lover's union, *marutam* with lover's infidelity, *neytal* with anxiety and *palai* with separation – and may be evoked by one or more elements characteristic of the landscape, such as a specific time, season, bird, animal, plant and occupation (Ramanujan 1967; Selby 2011). *Puram tinai* are named for situations such as prelude to war (*vetci*), beginning of invasion (*vanci*), siege (*ulinai*), battle (*tumpai*), victory (*vakai*), endurance (*kanci*) and elegy (*patan*); scholars, however, agree that, unlike the *akam* poems, correspondences between the landscape and the elements of the poem are not always clear in *puram* poems (Ramanujan 1985; Hart 1999).

4.5 Bards and Monkeys

Birds and animals are important components of the landscape in *akam* poems. Particular species symbolise certain landscapes, for example, the monkey, elephant, horse and bull typify the *kurinci* landscape, while the crocodile and shark identify the *neytal* landscape. However, 'birds and beasts of one landscape' are sometimes permitted to appear in other landscapes (Ramanujan 1967, 1985). Apart from the sophistication of the poetics, *Sangam* poems are also a historical record of people's knowledge and perceptions of animals and plants thousands of years ago and, for this reason, offer a fascinating field of investigation for naturalists and students of animal behaviour. Scholars who have examined the treatment of nature in *Sangam* poetry attest that details regarding plants and animals are portrayed with amazing accuracy in the texts (Varadarajan 1957; Thani Nayagam 1966). Imagery involving plants and animals are built on their natural properties, so that 'the real world (is) always kept in sight and included in the symbolic' (Ramanujan 1985).

What does classical Tamil poetry tell us about macaques, the focus of this volume? Quite a bit, as it turns out. Monkeys, along with boars, tigers and elephants, are popular entities in *akam* poems. In fact, the *Ainkurunuru* has a whole section devoted to monkeys called *Kurakku-p-pattu*, or Ten Poems on the Monkey (Selby 2011). In this segment, the poet uses the figure of a male monkey to refer to the hero and his behaviour, which is sometimes less than desired. The ten poems traverse the circumstances of the hero's love for the heroine, their attempts to meet clandestinely, the heroine's anxious wait for her lover and finally, their elopement and marriage (Selby 2011). Given below are a few selections from three of the poems:

In his land,
a foolish, strong monkey,
the young of a black-fingered female,
disturbs a comb full of sweet honey
on the treacherous mountain,
then springs onto a nearby branch,
long and unsteady.[13]

In his country,
the mate of the female monkey,
a fool of a male,
scampers up the rising slope of the hill
at the roaring of the mighty tigress[14]

In your country,
the lover of the female monkey,
that male who grazes on shoots,
takes up a cool, fragrant creeper
and slashes at young clouds
foaming over broad slabs of rock[15]

As the lines depict, the monkey epitomises the hero, who may be strong but is inept at stealing the 'sweet honey', caring yet thoughtless of his lover's feelings and playful but forgetful of her pain during his long absences (Selby 2011). The animal imagery used lends a tone of light-heartedness to the segment; although the heroine's friend chides the hero for his inconsideration and the heroine is saddened by his neglect, there is no real sense of misery or woe. In contrast, in *Kelar-pattu*, or Ten Poems on the Boar, where the boar is used to refer to the hero and his actions, the mood evoked is one of disquiet:

In his country,
a brave boar,
gorging on tender millet,
sleeps on a slope
strewn with hard stones.[16]

In his hilly country,
an enraged male boar
with tiny eyes
diverts the archers
near the plinth

[13] *Ainkurunuru* 272. From *Tamil Love poetry: The five hundred short poems of the Ainkurunuru (2011).* Translated by M.A Selby. New York: Columbia University Press.

[14] *Ainkurunuru* 274. From *Tamil Love poetry: The five hundred short poems of the Ainkurunuru (2011).* Translated by M.A Selby. New York: Columbia University Press.

[15] *Ainkurunuru* 276. From *Tamil Love poetry: The five hundred short poems of the Ainkurunuru (2011).* Translated by M.A Selby. New York: Columbia University Press.

[16] *Ainkurunuru* 261. From *Tamil Love poetry: The five hundred short poems of the Ainkurunuru (2011).* Translated by M.A Selby. New York: Columbia University Press.

on the slopes
then steals the paddy.[17]

The boar may be more stately and adept than the monkey, but he is untrustworthy in a way that the monkey is not. In the world of Sangam poetry, animals and 'people of low culture' only possess the five senses of touch, taste, smell, sight and hearing. As they lack the sixth sense of the mind, they cannot be expected to honour or cherish the higher values of life. Hence, when human characters behave thoughtlessly, their behaviour suggests parallels in the animal world (Ramanujan 1985; Selby 2011). It is notable that the Sangam poets had very specific views about the particular qualities of various animals. The monkey and the boar very clearly symbolise different sets of attributes, and as the poetry fragments below demonstrate, the monkey is consistently depicted as a denizen of the forests, daring, curious, greedy, mischievous and easily distracted:

that girl with the hair
that can't yet be tied in a knot
has crossed over those tangled wastes
unknown even to monkeys.[18]

And the bandits, men who steal like monkeys,
fierce and greedy and spread throughout
the coolness of the mountain dense with trees,
they are one more enemy, oblivious to our state[19]

Excited by such teeming voices,
an audience of female monkeys
watches in wonder
the peacock in the bamboo hill[20]

He is from those mountains
where the little black-faced monkey,
playing in the sun,
rolls the wild peacock's eggs
on the rocks.[21]

Frodsham (1967), in his comparative account of landscape poetry in Europe and China, casually, and most memorably, dismisses Sangam poetry for its 'limited appreciation of nature'. It may be useful to reflect on Frodsham's criticism in order

---

[17] *Ainkurunuru* 267. From *Tamil Love poetry: The five hundred short poems of the Ainkurunuru (2011).* Translated by M.A Selby. New York: Columbia University Press.

[18] *Ainkurunuru* 374. From *Tamil Love poetry: The five hundred short poems of the Ainkurunuru (2011).* Translated by M.A Selby. New York: Columbia University Press.

[19] *Purananuru* 136. From *The Purananuru: Four hundred songs of war and wisdom (1999).* Translated and edited by G. L. Hart and H. Heifetz. New York: Columbia University Press.

[20] *Akananuru* 82. From *Poems of Love and War: From the eight anthologies and the ten long poems of classical Tamil* (1985). Translated by A.K. Ramanujan. New York: Columbia University Press.

[21] *Kuruntokai* 38. *From Poems of Love and War: From the eight anthologies and the ten long poems of classical Tamil* (1985). Translated by A.K. Ramanujan. New York: Columbia University Press.

to analyse the usage of nature in Sangam poetry. Classical Tamil poetry does not display the mystical fascination for wild nature that is seen in early Chinese and English literature, which Frodsham defines as true landscape poetry. Instead, nature is part of the human landscape here. Animals, birds and plants are part of the human experience; they add to it, embody it and further refine it. In Ramanujan's (1985) words, 'mere nature description or imagism in poetry would be uninteresting to classical Tamil poets and critics, for it would not "signify"; it would be a signifier without a signified, a landscape (*mutal* and *karu*) without an *uri*, an appropriate human mood'.

Of more interest to the biologist is, of course, another set of concerns: What is the species identity of the monkey described in Sangam literature? Are they accurate representations of the species? Do the poems add anything to our corpus of knowledge about the species? Four species of anthropoid primates are found in southern India – the bonnet macaque *Macaca radiata*, lion-tailed macaque *Macaca silenus*, Hanuman langur *Semnopithecus entellus* and the Nilgiri langur *Trachypithecus johnii*. (Southern India is also home to a fifth primate species, the nocturnal prosimian slender loris *Loris tardigradus*, but for obvious reasons, this species is not part of the discussion here.) The bonnet macaque and the Hanuman langur are distributed across entire peninsular India while the lion-tailed macaque and the Nilgiri langur are restricted to the wet forests of the Western Ghats mountain range (Fig. 4.1). How does the distributional range of the four primate species map onto the geographical world of the Sangam poets?

'Tamilakam' or the Land of Tamils is described in Sangam poetry as a collection of small chiefdoms in southern India dominated by the three kingdoms of Pandyas, Cheras and Cholas (Sastri 1955; Champakalakshmi 1987; Selby and Peterson 2008). Three major river valleys were the loci around which these kingdoms flourished – the Pandyas in the Vaigai and Tamaraparani river valleys, the Cheras in the Periyar river valley and the Cholas around the lower Kaveri river basin (Stein 1977; Champakalakshmi 1987). Hence, significant parts of the distribution ranges of the bonnet macaque, lion-tailed macaque, Hanuman langur and Nilgiri langur must have been within the geographical boundaries of the region known to the Sangam poets (Fig. 4.1). The poems use several terms for monkeys: *ukam*, *mandhi*, *kaduvan*, *kalai*, *musu* and *kurangu*. *Kurangu* and *ukam* are generic terms for a monkey, while *mandhi* means a female monkey and *kalai, kaduvan* and *musu* refer to a male monkey. The word *parpu* is used to mean the infants of animals; when used in the context of a monkey description, it refers specifically to an infant monkey.[22] Although no specific terms are used for different kinds of monkeys, some of the poems describe a red-faced monkey and a black-faced monkey:

> Lord of a mountain where the slopes rise so high that the summit
> cannot be touched by the clouds and a male monkey with black fingers

[22] See http://sangampoemsinenglish.wordpress.com/nature-in-sangam-tamil/ and http://animalsinsangamtamil.wordpress.com for information on classical Tamil terms for various animals and birds and a list of Sangam Tamil poems that involve animal and bird descriptions.

Fig. 4.1 Map of southern India depicting the Eastern and Western Ghats mountain ranges and the main rivers

plucks and eats fruit from a green-leafed jackfruit tree towering
upon that mountain, handsome with his red-faced mate, a sight
glowing
into the distances and then he settles to sleep at the top of a bamboo![23]

Like a male black-faced monkey
who looks as if he's wearing face paint,
crashing through the forest
grabbing wildly at the branches
because he can't tell which
can support his weight[24]

---

[23] *Purananuru* 200. From *The Purananuru: Four hundred songs of war and wisdom (1999).* Translated and edited by G. L. Hart and H. Heifetz. New York: Columbia University Press.

[24] *Kuruntokai* 121. From *Kuruntogai* (2010). Translated by R Butler. Ebook.

While 'red-faced' certainly refers to the bonnet macaque (sexually cycling, lactating and some old female bonnet macaques acquire a characteristic red face; Sinha 2001), black-faced could refer to any of the other three primate species, the lion-tailed macaque, Nilgiri langur and Hanuman langur. However, the particulars of the description in the above lines, 'who looks as if he's wearing face paint', lead one to infer that the poet is talking about the Hanuman langur. Most of the poems do not carry such descriptors though; it would appear that the Sangam poets, except for noting the morphological variation, did not really distinguish between the different monkey species. The generalities of monkey behaviour are infinitely of greater interest to them; even the simplest image is created with a wealth of detail that says much for their observations of animal life:

It would be so good,
if only there were someone to hear
your complaint,
and hold you tenderly
like a young monkey
holds onto her mother[25]

when the clouds rumble
and the rain falls in sheets
a male monkey
his coat dense and bristling
reaches for a ripe jack-fruit
with its intense flower-like perfume
and sends it splashing down[26]

that male monkey
roaming the hillside
with his sturdy young son -
showing his red mouth
and thorn-like fangs
as he eats the ripe mango fruits[27]

A refrain that is associated with many of the monkey allusions refer to the monkey's fondness for jackfruit (*Akananuru* 2; *Kuruntokai* 90, 153; *Natrinai* 373; *Purananuru* 128, 200). One of the verses in *Kuruntokai* (342) even talks about how men need tie nets around jackfruit trees to protect the ripe fruits from the marauding monkeys. The image of the monkey as an opportunistic crop raider is reinforced in *Kuruntokai* 335, where monkeys grab the grain, left drying on the rocks, when no one is watching:

where girls wear fine jewellery
with bangles stacked upon their arms
spread red millet over the flat boulders,
which green-eyed female monkeys
with their young

[25] *Kuruntokai* 29. From *Kuruntokai: An anthology of classical Tamil love poetry (1976).* Translated by M. Shanmugam Pillai and David E. Ludden. Madurai: Koodal Publishers.

[26] *Kuruntokai* 90. From *Kuruntogai* (2010). Translated by R Butler. Ebook.

[27] *Kuruntokai* 26. From *Kuruntogai* (2010). Translated by R Butler. Ebook.

grab and run off with,
jumping down from the branches
picking a time
when the girls are splashing in the tank
and paying no attention to it.[28]

Although the Sangam poets were not interested in identifying the primate species they talked about, several points argue that it is indeed the bonnet macaque that they were most familiar with. For one, most of the poems (*Ainkurunuru* 272, 279; *Kuruntokai* 26; *Purananuru* 116) talk about leaping, playing monkeys, a description that sits better on the rambunctious macaques than the stately langurs. The first verse of *Kurakku-p-pattu* in *Ainkurunuru* (271) talks of how when the female monkey eats *avarai* beans, her cheeks resemble the full bags of merchants[29] – a graphic simile that immediately evokes an image of the stuffed cheek-pouches of bonnet macaques. Then there is the striking account of a performing monkey infant, 'fig-coloured' and 'red-faced' that balances on a high rope, watched by *kuravar* children (*Natrinai* 95).[30] The description is clearly that of a bonnet macaque; more interesting, even today, the *kuravars*, a nomadic tribe known for their interest in hunting and trapping wild animals, use bonnet macaques as performing pets in Tamil Nadu (Meshack and Griffin 2002; Peterson 2008). The portrayal of the bonnet macaque as a pet recurs in *Natrinai* 353; here, the 'the playful daughter' of the *kuravan* is said to be 'feeding jackfruit to a black-fingered monkey'.[31] Most striking, perhaps, are depictions of the monkey living close to human settlements and found even in the backyard of one's house, evidence that even 2,000 years ago, the bonnet macaque was behaving in a very weed macaque-like fashion:

Once: if an owl hooted on the hill,
if a male ape leaped and loped
out there on the jackfruit bough in our yard
my poor heart would melt for fear.[32]

Macaques are inquisitive, troublesome, grasping and lively in Sangam poetry, but quite remarkably, they are not holy or godly. The deification of monkeys was a much later process, expressed most decidedly in the form of righteous Hanuman in *Iramavatharam*, or the Tamil *Ramayana* (circa 9–12 CE) by the poet Kamban. In the Sangam landscape, macaques are just there, much like Mallory's Mount Everest,

[28] *Kuruntokai* 335. From *Kuruntogai* (2010). Translated by R Butler. Ebook.

[29] See https://animalsinsangamtamil.wordpress.com for this translation. For an alternate translation of the same verse, see Selby (2011): "a female monkey…her stomach swelling/like a peddler's sack". A strictly literal translation of the original verse would be "the female monkey appears like …".

[30] https://animalsinsangamtamil.wordpress.com.

[31] *Natrinai* 353. From Love Stands Alone: Selections from Tamil Sangam poetry (2010). Translated by M.L. Thangappa. Edited by A.R. Venkatachalapathy. New Delhi: Viking Penguin.

[32] *Kuruntokai* 153. From *The Interior Landscape: Love poems from a classical Tamil anthology* (1967). Translated by A.K. Ramanujan. Bloomington: Indiana University Press.

and it is indeed a pity that we have come so far that we can no longer empathise with the equanimity of the poet Mutamociyar when he sang of the monkey in the town's commons[33]:

When the ape
on the bough
of the jackfruit tree
in the town's commons
mistakes for fruit
the eye
on the thonged drumheads
hung up there by mendicant bards,
he taps on it,[34]

Acknowledgements I am extremely grateful to Martha Ann Selby, Latha Raman and Anindya Sinha for reviewing this chapter; their perceptive comments helped to significantly improve an earlier version of this chapter.

References

Alinei M (1997) Magico-religious motivations in European dialects: a contribution to archaeolinguistics. Dialectologia et Geolinguistica 5:3–30

Bacon F (1960) The new organon and related writings. Liberal Arts Press, New York (Originally published in 1620)

Baldick J (2000) Animal and shaman: ancient religions of central Asia. I B Tauris & Co. Ltd., London

Barnett SA (1990) Animals, analogy and models of humanity. In: Tapia LL (ed) Para conocer al hombre: Homenaje a Santiago Genoves. Universidad Nacional Autónoma de México, Mexico, pp 41–48

Barron P (2002) The separation of wild animal nature and human nature in Gilgamesh: roots of a contemporary theme. Pap Lang Lit 38:377–394

Briggs WW (1980) Narrative and simile from the Georgics in the Aenid. Brill, Leiden

Burghardt GM (1985) Animal awareness: current perceptions and historical perspective. Am Psychol 40:905–919

Carporael LR, Heyes CM (1997) Why anthropomorphize? Folk psychology and other stories. In: Mitchell RW, Thompson NS, Miles HL (eds) Anthropomorphism, anecdotes, and animals. State University of New York Press, Albany, pp 59–76

Champakalakshmi R (1987) Urbanisation in south India: the role of ideology and polity. Soc Sci 15:67–117

Cook M (1999) Ibn Qutayba and the monkeys. Stud Islam 89:43–74

Doniger W (2005) Zoomorphism in ancient India: humans more bestial than beasts. In: Daston L, Mitman G (eds) Thinking with animals: new perspectives on anthropomorphism. Columbia University Press, New York, pp 17–36

[33] *Purananuru* 128. From *Poems of Love and War: From the eight anthologies and the ten long poems of classical Tamil* (1985). Translated by A.K. Ramanujan. New York: Columbia University Press.

[34] This verse was, in many ways, my real introduction to the world of classical Tamil poetry, and I am indebted to Anindya Sinha for awakening me to this thing of beauty.

Fisher JA (1991) Disambiguating anthropomorphism: an interdisciplinary review. In: Bateson G, Klopfer PH (eds) Perspectives in ethology, vol 9. Plenum, New York, pp 49–85
Frobenius L (1929) Monumenta terrarum: Der geist uber den Erdteilen. Buchverlag, Frankfurt
Frodsham JD (1967) Landscape poetry in China and Europe. Comp Lit 19:193–215
Gale MR (2000) Virgil on the nature of things: the Georgics, Lucretius and the didactic tradition. Cambridge University Press, Cambridge
Grant JN (1969) Dido Melissa. Phoenix 23:380–391
Hart GL (1999) Introduction. In: Hart GL, Heifetz H (trans and eds) The Purananuru: four hundred songs of war and wisdom. Columbia University Press, New York, pp 15–37
Kennedy JS (1992) The new anthropomorphism. Cambridge University Press, New York
Knight J (1999) Monkeys on the move: the natural symbolism of people-macaque conflict in Japan. J Asian Stud 58:622–647
Kovacks MG (trans) (1989) The Epic of Gilgamesh, Stranford University Press, California
Layton R (2000) Shamanism, totemism and rock art: Les chamanes de la préhistoire in the context of rock art research. Camb Archaeol J 10:169–186
Levi-Strauss C (1962) Totemism. Beacon, Boston
Levi-Strauss C (1963) The bear and the barber. J R Anthropol Inst Great Britain Ireland 93:1–11
Lutgendorf P (2007) Hanuman's tale: the messages of a divine monkey. Oxford University Press, New York
Meshack D, Griffin C (2002) Displacement and forced settlement: gypsies in Tamil Nadu. In: Chatty D, Colchester M (eds) Conservation and mobile indigenous peoples: displacement, forced settlement and sustainable development. Berghahn Books, New York, pp 261–276
Mitchell RW, Thompson NS, Miles HL (1997) Taking anthropomorphism and anecdotes seriously. In: Mitchell RW, Thompson NS, Miles HL (eds) Anthropomorphism, anecdotes, and animals. State University of New York Press, Albany, pp 3–11
Olivelle P (1997) Introduction. In: The Panchatantra: the book of India's folk wisdom (trans: Olivelle P). Oxford University Press, Oxford.
Peterson IV (2008) The drama of the Kuravanci fortune-teller: land, landscape and social relations in an eighteenth century Tamil genre. In: Selby MA, Peterson IV (eds) Tamil geographies: cultural constructions of space and place in south India. State University of New York Press, Albany, pp 59–86
Pierce DC (1969) The middle way of the Jātaka tales. J Am Folk 82:245–254
Radhakrishna S, Sinha A (2011a) Dr Jekyll and Mr Hyde: the strange case of human-macaque interactions in India. Curr Conserv 4:39–40
Radhakrishna S, Sinha A (2011b) Less than wild: commensal primates and wildlife conservation. J Biosci 36:749–753
Ramanujan AK (1967) The interior landscape: love poems from a classical Tamil anthology. Indiana University Press, Bloomington
Ramanujan AK (1985) Poems of love and war: from the eight anthologies and the ten long poems of classical Tamil. Columbia University Press, New York
Ramanujan AK (1992) Three hundred *Ramayanas*: five examples and three thoughts on translation. In: Richman R (ed) Many Ramayanas: the diversity of a narrative tradition in south Asia. Oxford University Press, New Delhi, pp 22–49
Ryder AW (1925) Translator's introduction. In: The Panchatantra (trans: Ryder AW). The University of Chicago Press, Chicago
Saraswat R (2010) Rhesus macaque-human conflict in northern India: attitudes and perceptions. MSc Dissertation, TERI University, New Delhi
Sastri KAN (1955) A history of south India: from prehistoric times to the fall of Vijayanagar. Oxford University Press, New Delhi
Selby MA (2011) Tamil love poetry: the five hundred short poems of the Ainkurunuru. Columbia University Press, New York
Selby MA, Peterson IV (2008) Introduction. In: Selby MA, Peterson IV (eds) Tamil geographies: cultural constructions of space and place in south India. State University of New York Press, Albany, pp 1–16

Singh M, Rao N (2004) Population dynamics and conservation of commensal bonnet macaques. Int J Primatol 25:847–859
Sinha A (2001) The monkey in the town's commons: a natural history of the Indian bonnet macaque. National Institute of Advanced Studies, Bangalore
Smith BK (1991) Classifying Animals and Humans in Ancient India Man, New Series, 26: 527–548
Stein B (1977) Circulation and the historical geography of Tamil country. J Asian Stud 37:7–26
Sterckx R (2002) The animal and the daemon in early China. State University of New York Press, Albany
Sterckx R (2005) Animal classification in ancient China. East Asian Sci Technol Med 23:26–53
Thani Nayagam XS (1966) Landscape and poetry: a study of nature in classical Tamil poetry. Asia Publishing House, Bombay (currently Mumbai)
Varadarajan M (1957) The treatment of nature in Sangam literature. South India Saiva Siddhanta Works Publishing Society, Madras (currently Chennai)
Viereck W (2002) Insights into the cultural history of Europe as provided by the *Atlas Linguarum Europae*. Stud Angl Posnan 37:3–15
Wheatley BP, Harya Putra DK (1995) Hanuman, the monkey god, leads conservation efforts in the Balinese monkey forest at Ubud, Indonesia. Primate Rep 41:55–64
White LT Jr (1967) The historical roots of our ecologic crisis. Science 55:1203–1207
Winkelman M (2002) Shamanism as neurotheology and evolutionary psychology. Am Behav Sci 45:1873–1885
Winkelman M (2004) Shamanism as the original neurotheology. Zygon 39:193–217
Zvelebil K (1973) The smile of Murugan: on the Tamil literature of south India. Brill, Leiden

Part III
Cooperative Relationships Between Humans and Macaques

A monkey lived in a forest. Once, as he leapt from tree to tree, a splinter broke in his tail. He was sitting on a branch howling in pain and nursing his tail when he saw a barber walking that way. He called the man closer and begged of him: "Brother Barber, Brother Barber, a thorn or something is hurting my tail. Please take it out with your razor, bless you."

When the barber was taking out the splinter, which was in a tricky place, the whole tail broke off with a snap. The monkey began to cry in a loud voice, "Give me my tail back! My tail, my tail!"

The barber didn't know what to do. So he said, "Take my razor instead of your tail."

—*An Indian folktale*[1]

[1]Ramanujan AK (1997) A flowering tree and other oral tales from India. Penguin Books, New Delhi, p 109

Chapter 5
Macaques and Biomedicine: Notes on Decolonization, Polio, and Changing Representations of Indian Rhesus in the United States, 1930–1960

Neel Ahuja

5.1 Introduction

The current institutional framework for state-sponsored biomedical primate research in the United States was launched in 1960 with the opening of the Oregon Regional Primate Research Center. The Oregon center was the first of eight primate research centers funded by the National Institutes of Health (NIH), which in the second half of the twentieth century would make the breeding, confinement, and standardization of rhesus macaques its primary means for supplying US biomedical researchers with experimental animal models for human disease. The eight primate centers supply rhesus macaques with the specific pathogen-free (SPF) designation as well as small numbers of chimpanzees and vervets for applied research. These institutions have played important roles in a number of biomedical watersheds, including the isolation of HIV, the extraction of stem cells, and the first successful attempt at primate cloning.

In the reshuffling of Centers for Disease Control (CDC) and NIH departments that occurred following the September 11, 2001 attacks, the eight primate centers were renamed National Primate Research Centers, emphasizing the central role that nonhuman primate bodies have played in the state-corporate-university complex's engineering of US American immunity. In its current incarnation, the US biosecurity apparatus imagines the rhesus macaque as an integral species for the securitization of the nation and is often posed as a defense against a future of unknowable transnational diseases and health risks for which an ongoing

N. Ahuja (✉)
Department of English and Comparative Literature,
University of North Carolina-Chapel Hill, Chapel Hill, NC 27599-3520, USA
e-mail: nahuja@email.unc.edu

S. Radhakrishna et al. (eds.), *The Macaque Connection: Cooperation and Conflict between Humans and Macaques*, Developments in Primatology: Progress and Prospects 43,
DOI 10.1007/978-1-4614-3967-7_5, © Springer Science+Business Media, LLC 2013

diversification of biomedical research is necessary.[1] Yet if the nominal nationalization of rhesus labs seems tied to a specific and recent set of transformations of the US state in response to the specter of terrorism, the history of the primate centers' formation actually demonstrates their long-standing imbrication in the politics and imagined possibilities of national defense emerging from the Cold War. It was during the US polio scare of 1955, when the mass production of polio vaccine created a demand for hundreds of thousands of rhesus macaques, that national officials in charge of marshaling "research resources" established the rhesus as the primary biomedical model for the human to be imported and preserved as a vital national resource.[2] Coincident with the emergence of biomedical metaphors that figured the body as a site of battle waged by an internal immune system to contain the threats of the external world, the new national strategy emphasized the modeling of human biologic systems on the bodies of experimental monkeys figured as conscripts in an immunological battle for national security (Martin 1994). The discussions and institutional experiments leading up to the establishment of the primate centers document a lengthy process whereby US researchers attempted to gain access to nonhuman primate bodies in the Caribbean, central Africa, and South Asia. When polio epidemics spurred a high demand for Indian-origin rhesus, the recent and ongoing successes of nationalist movements in anticolonial struggles across countries with nonhuman primate populations brought about a change in these researchers' understandings of the proper institutional organization for primate research and of rhesus macaques themselves. In turn, scientific institutions were able to mobilize new conceptions of primate bodies in public discourse, which led to an increasing association of monkeys and apes with projects of US American technoscientific modernity.

In this chapter, I explore the changing representations of rhesus macaques and other nonhuman laboratory primates in the United States from 1930 to 1960, a period during which I argue that the increased visibility of the rhesus (as well as the popularization of "the research monkey" in visual culture) coincided with a process in which nonhuman primates were culturally nationalized or domesticated. While in the 1930s monkeys, chimpanzees, and gorillas had typically been figured as representing the untamable nature of animality and were associated with mad scientists and the supposed danger of the colonial jungle, the use of rhesus and other primates by biomedical researchers contributed to a new view of monkeys and apes as kin of humanity. This in turn made nonhuman primates possible conscripts in Cold War biomedical projects for national defense. This set of transformations cannot be understood as an example of scientific knowledge "modernizing" society by establishing an objective view of other species. While rhesus were like rats and fruit flies in that

[1]This future-oriented approach to health and security in the rich countries is described by Andrew Lakoff as "global health security," an assemblage that attempts to insulate wealthy countries from the presumed dangers of disease spread from Europe's former colonies (Lakoff 2010). A wide array of research on US imperial public health and medicine documents the techniques of US immunitary engineering against imperial frontiers and transnationally constituted disease threats (Merkel 1997; Shah 2001; Stern 2005; Anderson 2006; Wald 2008).

[2]For more on polio research agendas, see Paul 1971; Rogers 1992; Wilson 1998; Shell 2005.

they underwent important processes of standardization for diverse forms of laboratory labor in the twentieth century (Clause 1993; Kohler 1994; Clarke 1998), the shift in the social imaginary of the rhesus and other nonhuman primates was born out of the geopolitical and biomedical realities specific to these species' imbrication in a transnational politics of neocolonial research resource extraction.

This politics required the ultimate move of rhesus out of free-ranging habitats and into captive laboratory environments. While early primate researchers often shared the colonial assumption that nonhuman primates were best bred and studied in the supposedly "natural" environments of their African and Asian habitats, it was only after the wave of nationalist movements reestablished home sovereignty over formerly colonized ecosystems that US researchers attempted the type of broad-scale importation and breeding of rhesus culminating in the National Primate Research Centers. Decolonization brought about the "modern" institutions of US American primate laboratories, allowing the rhesus to be domesticated in three senses: rhesus macaques were literally imported into the continental boundaries of the nation; they were broken of their cultural association with a wild and untamed animality; and they were increasingly located in captive homes that attempted to scientifically and affectively establish their suitability to model human bodies and minds.[3] These processes of domestication undid the popular association of medical research with the history of colonial animal expeditions that had historically made primate research possible. Thus, as the US began to accept nonhuman primates as kin and as Cold War conscripts, images of rhesus macaques and other laboratory primates continued to reinforce a conceptual division between First and Third Worlds by demonstrating that the modern laboratory could make these previously sensationalized animals docile.[4] Ironically, as decolonization brought macaque breeding and experimentation within the national borders, this move worked to replicate a type of "domestic containment" that had posed the US American home/land as a space securitized against a communism figured as political, moral, and nuclear contagion (May 1990; Nadel 1995). The domestication of the rhesus macaque thus contributed to an American exceptionalist denial of US empire in the history of US biomedicine (Kaplan 1993)—a history that this chapter brings to light by tracing the

[3]Animal studies theorists have deployed a critique of the wild/domestic binary (Russell 2002), emphasizing the diversity of human-animal relationships crossing biological and social phenomena. I retain the term "domestication" in order to suggest the ideological linkage of rhesus to the racialized and gendered spaces of nation, family, and home.

[4]I draw here on the biopolitical theory of Michel Foucault (1995), who suggested that the eighteenth century saw the rise of a "power over life" or biopower emphasizing the institutional production of "docile bodies" which were evaluated and optimized using medical knowledges. More recently, the interdisciplinary fields of animal studies (Wolfe 2010; Shukin 2009; Wadiwel 2002) and science studies (Rose 2007; Latour 1993) have built on Foucault's theory in the contexts of transpecies and transgenic forms of contemporary biomedicine and agriculture. The making of "docile bodies" further suggests the implication of biopower in the production of racial power, as race played an important role in understanding which bodies were made fit or docile for national uplift (Foucault 2003; Stoler 1995).

writings and travels of US primatologists alongside broader public representations of the transpecies kinship of human and macaque.

5.2 A Tropical Rhesus Colony: The Puerto Rican Experiment

The story of the transformation of the rhesus into the standard biomedical model for the human begins in the 1930s. At a time when South Asian humans were denied citizenship in the United States and largely restricted from immigrating, the Columbia University comparative anatomist Clarence Ray Carpenter imported Indian monkeys as supreme scientific models for the human body. Carpenter, who helped spur a contested project of a medical school run by Columbia at the University of Puerto Rico-San Juan, spearheaded the organization of the first free-ranging rhesus colony on US-controlled land when he brought 409 Indian rhesus to Cayo Santiago, Puerto Rico. This islet in the Puerto Rican archipelago had been a sugar-exporting platform during the years of Spanish rule and into the early decades of its status as a US American commonwealth. The 1939 *Life* magazine article introducing the Indian-origin monkeys at Cayo Santiago initially focused on familiar colonial tropes exoticizing India and defending British colonialism with reference to the presumed irrationality of Hindu spiritual practices: "Because he is considered sacred in India," claimed the unnamed writer, "the rhesus is domineering, undisciplined and bad tempered" (Anonymous 1939). Following the logic in the *Life* article, many of the popular and scientific representations of rhesus in the years preceding the World War II rhetorically linked rhesus macaques to the discourse of the jungle and the narratives of colonial expedition that framed many early primate researchers' encounters with their research subjects. Carpenter's experiment in creating a macaque colony reflects the dominant assumption that free-ranging primate colonies would reproduce a "natural" tropical environment in which rhesus behavior and social organization would be best preserved.

Carpenter brought Indian rhesus to Puerto Rico, which had been constructed as a natural and ideal space of primate reproduction—a place which US scientists had viewed as an enclosed and exploitable ground for collecting data and specimens since the early days of US occupation (Duany 2002). Nearby residents in Punta Santiago, Puerto Rico, were "alarmed" by the *Life* article's mention of planned use of the colony to research cures for polio and other infectious diseases (Rawlins and Kessler 1986). A community group met with colony scientists to voice their concerns about risk of disease transmission to humans in the area. Although this initial public resistance to the colony apparently died down after a forum attended by scientists, the history of Cayo Santiago's founding testifies to the imbrication of biomedicine within histories of imperial power. Utilizing the institutional resources of tropical medicine, an association of US science with Puerto Rican modernization discourse, the labor of primate traders in British India, partnerships with the US military, and the decaying infrastructure of the sugar trade, the Cayo Santiago colony represents a trajectory of transcolonial relations whereby the impacts of war, increased transnational contact, and medical science led researchers across oceans for the harvesting of rhesus macaques as raw biological materials.

Carpenter's reasons for establishing the Cayo Santiago colony were twofold: (1) the colony would offer a stable field environment for behavioral studies of rhesus macaques who were relatively undisturbed by humans, and (2) it would provide laboratory scientists with a stable and healthy source of subjects for experimentation and vaccine production. A number of primate colonies had been established around the world in the 1920s, but none fulfilled Carpenter's vision of providing nonhuman primates for both lab and field study for US researchers. The earlier colonies, located in Cuba, Tunisia, French Guinea, Tenerife, Georgia, and elsewhere, were usually established in tropical medicine institutes of colonial powers or as philanthropist-funded initiatives. "Primate studies" of the 1920s "were a colonial affair, in which knowledge of the living and dead bodies of monkeys was part of the system of unequal exchange of extractive colonialism" (Haraway 1989, p.19). As an emissary of one of the US's top research universities, Carpenter was poised to both harness institutional resources and to fashion a global strategy for American research resources, combing the expanding global frontiers of US empire stretching from Asia to the Caribbean in order to ensure researcher access to nonhuman primate bodies.

In 1937, Carpenter traveled across Asia as part of the Asiatic Primate Expedition, which included renowned primate researchers Harold Coolidge, Adolph Hans Schultz, and Sherwood Washburn. These researchers convened in Singapore and fanned out to other sites, including China, North Borneo, and Java. Spurred by disruptions in the primate trade due to the First World War, as well as the increased demand for research subjects in the US, the expedition simultaneously sought out research subjects and provided an actual vehicle for US defense intelligence. A retired US defense intelligence agent recounts that the expedition brought along at least one undercover US operative to station in China to monitor Japanese activities (Noble 2006).

In 1938, the year Carpenter traveled to India to harvest macaques for Cayo Santiago, the United States imported nearly 16,000 rhesus macaques for scientific experimentation. Yet the situation in colonial India—the source of the most prized specimens for research—was less than ideal for Carpenter and other US scientists. The Indian rhesus, known for its ease of captivity, its close association to human anatomy and metabolism, and initial easy accessibility under British colonialism, became in the 1930s a target of animal advocacy organizing. Indian animal advocates including the SPCA pressured the British Indian government to institute its first rhesus protection regulations in 1937. These regulations banned the transport of rhesus during the summer months when high temperatures in holding pens on trains and ships frequently caused high rhesus mortality rates (Carpenter 1940). Additionally, scientists reported wartime disruptions in shipping of primates.

Carpenter told the story of his own work to secure a steady supply of rhesus within the rhetorical frame of the colonial expedition narrative. Such stories of colonial adventure and scientific or capital exploitation were immensely popular in the 1930s, which saw a resurgence of expedition narratives in popular fiction and film.[5] The most popular of these stories—the series of *Tarzan* stories and the 1933 film classic *King Kong*—figured apes as both humanity's closest ancestors and their gravest danger in the colonial jungle. At this time and into the 1940s, the figure of the laboratory monkey (often actually an ape or chimpanzee, or else named interchangeably as monkey and ape) dominated the US imagination of both the laboratory and the colonial world. Visual depictions of monkeys and especially great apes loomed large in the national imaginary both due to the high visibility of apes featured in domestic circuses and the heightened racist stereotyping of the decade associating social change with the supposed animality and hypersexuality of African-American men (Bederman 1993). At the same time, and in line with colonialist language figuring the colonized as similar to "lower primates," the rhesus macaque was also often a source of comedy emphasizing monkeys' mimetic capacities. In fact, the first popular image of the rhesus macaque was Hansel Mieth's 1939 *Life* magazine image of a Cayo Santiago rhesus that had swum just off the coast of the islet. While Mieth saw the monkey's angry face as a sign of modern alienation, the *Life* editors made light of the monkey by claiming it had swum away from "the chatter of innumerable female monkeys."

Carpenter's own 1959 narrative of his macaque expedition figured the complexities of the Indian colonial social world as a key danger for the progress of modern biomedicine. His story at times reads like *King Kong*, as the unprepared masculine adventurer overcomes financial hardships to set out on "the very nervy business" of going across vast oceans to trap the primate specimen. Yet the possibility of an even modest heroism for Carpenter within his own narrative melts upon arrival in India, as Carpenter presents Indian disease, greed, communalism, and nationalism overwhelming his efforts at supplying science with macaques. Indian greed is the first obstacle presented in Carpenter's narrative. Being unable to personally collect the hundreds of specimens needed for his colony, Carpenter seeks the services of animal traders, who he later decides are part of a "worldwide racket" stretching from Calcutta to New York to New Orleans. The traders, according to Carpenter, made "fantastic charges for animals that are bought for practically nothing in their local area" in and around the city of Lucknow. Beneath the traders, writes Carpenter, served two distinct "animal unions" consisting of Muslim trappers and Hindu caretakers. The Muslims engaged in a "rough and cruel business" of trapping that contrasted with the hypersentimentality of the Hindus for whom the animals were "sacred" (Rawlins and Kessler 1986, p. 15–18). Reinforcing US and British popular conceptions of the

[5]Popularized at the height of salvage anthropology in the 1930s, the zoological expedition brought scientists, hunters, zookeepers, taxidermists, and filmmakers to various locations across the colonial world, searching for often elusive prize animals to be studied, hunted, and captured as a spectacle of the power of modern science (Rony 1996, p. 154 and 157–60). As such, the zoological expedition was one method of monetizing imperial power through what Timothy Brennan (2005, p. 101) calls the "economic image-function of the periphery": the operation of "the idea of the global periphery" as "an economic engine." In this case, the imperial gaze of the filmmaker, the scientist, and the zoological tourist are all connected to circuits of exchange, primarily founded on the existence of particular animal ranges in Africa and Asia.

colonial world as locations of disease bred by poverty, dirt, and overpopulation, Carpenter reports that the monkeys were exposed to the "disease-carrying humans" that "usually surrounded" the "filthy" Indian ports (Carpenter 1940, p. 285). Carpenter had each animal isolated and tested before the voyage.

In a 1940 article he published in *Science* on the primate trade between India and the US, Carpenter takes pains to critique the inhumane practices of trapping and transporting monkeys, but only in terms of the damage that cruel practices cause to scientific research. The exuberance with which the Society for the Prevention of Cruelty to Animals, Hindus, Buddhists, conservationists, and even the government of British India have attempted to limit the primate trade, according to Carpenter, is misguided as the monkeys are a "necessary import." Carpenter implies that corrupt nationalists manipulate Indians to support animal protection: "These peoples... are told that monkeys are used for the 'rejuvenation of decadent Westerners'" (Carpenter 1940, p. 285). The *Life* article on the opening of the colony, which cites Carpenter and animal caretaker Michael Tomilin as its only sources, claims that the monkey trade is in danger "because Mahatma Gandhi is preaching against the exportation of the sacred rhesus monkey" (Anonymous 1939). In response, Carpenter has faith that strong US diplomacy with the British Indian administration will secure scientists a steady supply of Indian monkeys.

Carpenter closes his narrative with a description of his journey by sea to Puerto Rico. His irritation over the logistical difficulties permeates this section of the narrative. Underfunded, Carpenter recounts shortchanging the captain of the ship who agreed to take his unusual cargo. Carpenter has no time to dwell on his disdain for Indians, sailors, and the poor given the grueling duties of tending to hundreds of animals, especially given that his rations quickly spoiled and he had to secure new food sources at a stop in Ceylon (Sri Lanka). Surviving the cold around the Cape of Good Hope, Carpenter headed for New York where he shifted the monkeys to an American merchant marine ship, the SS Coamo, bound for Puerto Rico. Carpenter recalls how "delighted" he was "to turn this shipment over to Mr. Michael Tomilin... or whoever else wanted to care for them" (Rawlins and Kessler 1986, p. 19). Carpenter abruptly ended his management of the colony and abruptly ends his narrative.

Yet while Carpenter could leave satisfied at having transported his 409 rhesus monkeys to Puerto Rico, other officials involved in the day-to-day operations of Cayo Santiago would find themselves charged with the task of producing a livable habitat for macaques. And instead of looking to the Indian habitat as a model, imperialist fantasies of a tropical nature filled with palms and coconuts (possibly fueled by US popular representations of its new possessions in the Caribbean and Pacific) came to structure the new landscape of Cayo Santiago. As Carpenter was away on his India voyage, Columbia officials were directing the transformation of the pastureland on the islet into a tropical paradise from which the monkeys could indefinitely feed. The New Deal Civilian Conservation Corps was enlisted to forest the island with mahogany, coconut palms, fruit trees, and root vegetables. In advance of the release of the monkeys, *The Illustrated London News* presented images and descriptions of the lush flora in travel-postcard prose: "It is... rocky land covered with shade-trees and thickets; some of it sandy depression, fringed with coconut-palms, and some flat, earthy expanses in which... root crops have been planted for the future islanders.... It is, in effect, a miniature Pitcairn Island dropped into a shallow sea of parrot-wing blue, jade green, or wrinkled copper, according to the caprice of the sun and wind" (Locke 1938, p. 290). Yet save a few coconut palms, most of these transplanted organisms were killed off by climate or hungry monkeys (Rawlins and Kessler 1986). Presuming that tropical fruit would nourish animals and that the sea would enclose them, caretakers were surprised when the monkeys destroyed the expensive fruit trees and, in small numbers, swam to the Puerto Rican main. Tearing down the imperialist fantasy of tropical nature, the animals forced scientists to establish feeding stations. This in turn had material and communicative effects, provisioning monkeys and training them to see built platforms and humans as part of social life. Cayo Santiago ultimately had to be maintained as a "semi-free-ranging" site, with feeding stations featuring Purina monkey food and cages for intermittent examinations becoming standard parts of life on the island. Colony employees and scientists learned not to disturb or provoke monkeys by wearing sunglasses or directly staring at the animals.

Despite the effort put into establishing the colony at Cayo Santiago, after Carpenter's 1940–1941 field studies, many of the initial caretakers left, and, with the Columbia-run School of Tropical Medicine taken over by the University of Puerto Rico, funding became scarce. Local residents had to feed the monkeys in the absence of state support. The colony was saved by the efforts of the Puerto Rican biologist José Guillermo Frontera who secured a grant from the National Institute of Neurological Diseases and Blindness, which carried out small studies in the 1950s. However, after a long gap in funding and with air transport speeding the pace of imports directly from India, researchers soon focused on building new, mainland US facilities. Cayo Santiago would, by the end of the 1950s, become an example for US research resources officials of a failed plan for building a national supply of research macaques. Instead, NIH officials would think carefully about both global primate sourcing and the possibilities of domestic primate breeding.

5.3 India, Central Africa, and the Cold War Push for a National Rhesus Macaque Strategy

During World War II, popular representations of medical science's encounter with nonhuman primates often consisted of highly sensational narratives of jungle expeditions to find miracle cures or tales of mad scientists corrupting human society through xenotransplantation between humans and apes. Such ideas are best expressed in films such as the 1940 production *The Ape*, which labeled the research scientist himself as an ape for losing touch with his own humanity. As a vivisectionist willing to sacrifice rats, then dogs, then apes, and finally humans, the polio researcher who used laboratory primates was portrayed as a flawed scientist who disturbed the species boundary in a futile attempt to extend life. In an even more sensational example, the 1945 serial *Queen of the Jungle* featured a US American girl traveling to central Africa and, after being separated from her group, becoming the white queen among a tribe of indigenous Africans. Her adopted community worships a giant ape as a jungle god protecting dangerous yet biomedically revolutionary uranium, which is later extracted along with the queen herself by a dashing American scientist. Wartime anxiety, as well as the much-publicized inability of researchers to fashion an experimental cure for polio over decades of primate research, required that futuristic promises regarding biomedical experimentation be qualified with the dangers that raw biomedical materials and colonial locations of biocapital extraction posed to the integrity of the white body and the individual's control over social life. Revealing an identification of the US cinematic gaze with British colonial conquest and expedition narratives (Shohat and Stam 1994), these dangers were often visually represented through a set of colonial tropes that made forested landscapes into a racial crucible—the large mammals of the jungle appeared as the masculinized, dark bodies of the so-called savage world.

Yet immediately following World War II, both the institutional structures and public representation of biomedicine's relationship to nonhuman primates shifted. The emergence of a medical consumer culture in the 1950s helped make US publics understand medical products as an essential component of "a prosperous economy and a modern social self" (Serlin 2004, p. 3). This medicalization of social life and US national identity emerged alongside US imperial exuberance over the nation's technological supremacy following the war. The atomic resolution to the war had both demonstrated the nation's technological prowess and had, according to state propaganda, helped the US avoid the mass destruction of a presumed future attack on the continental US by Japan. After the war's end, the government would harness the atomic "victory" as well as fear over heavy Soviet investment in technology in order to obtain significant state subsidies for research. State support for science and technology became heavily institutionalized, eclipsing the philanthropic research system of the wartime era. With increased funds for national health and medicine, researchers would seek out new and increasingly complex tools for fighting dread diseases like polio (Cook-Deegan and McGeary 2006).

At this time, after Carpenter had switched his professional focus to communications and worked as a filmmaker for the US war effort, several officials at NIH as well as in the Defense Department and the State Department began discussions of expanding Carpenter's original idea to promote a national system to ensure macaque populations for key research objectives (Garden 1992; CENCPC 1957). While these officials' interests were quite varied—ranging from an interest in radiological testing to the support of specializations such as blindness and heart disease—the effort to nationalize primate research did not begin to take shape until after a decade of discussions, by which time Tulane University heart researcher George E. Burch convened the Committee for the Establishment of a National Cardiovascular Primate Colony (CENCPC) at NIH. From 1956 to 1963, this committee would bring about the development of the national system for the importation and breeding of rhesus macaques. And in its initial phases, following the logics of Carpenter and the Asiatic

Primate Expedition, CENCPC was often focused on diversifying primate sources by establishing overseas sources and colonies for the study of nonhuman primates.

By the time Burch helped establish the first Regional Primate Research Center, the number of animals used in research had grown to immense proportions. In the early 1960s, the NIH reported that in addition to the four million animals used in its own labs, nine million were used in private pharmaceutical research. In addition, there were millions more uncounted animals used in hospitals, universities, and other research institutions. Nearly one-half of all research funded by NIH in 1963 used animal subjects (Animal Resources Branch 1963). Yet rhesus availability was exhaustible. Because of physiological differences between Indian rhesus macaques and members of the same species located in China, Indian imports were defined as a priority by US researchers. There were also various political and economic concerns, disease, and the problem of human encroachment, all of which limited the transfer of animals from their range in present-day Pakistan, Nepal, India, and Bangladesh. Imports during the mid-1950s were the highest ever, numbering around 150,000 annually, with 120,000 of that number killed for vaccine production. (In later years, more efficient vaccine production techniques cut that number nearly in half (Haraway 1989).) In the minutes of Burch's early meetings with other primatologists to formulate proposals for a national primate strategy in 1957, despite the large number of rhesus imports to the US, researchers noted the increasing difficulty in securing imports for smaller researchers and for specified types of animals. This situation meant that larger importers—especially the companies making polio vaccine—dominated the market. Ninety percent of imported monkeys, they estimated, were used for the production of polio vaccine. Leon Schmidt, a researcher at a Cincinnati hospital, claimed, "It looks like we have passed the point of availability of the larger animals and we're dipping into the young now who would be under normal circumstance the creators of the next generation. So it looks as if we are in a very, very precarious situation with respect to procurement of rhesus monkeys from the particular areas of India where they have been found in former years" (CENCPC 1957, p. 18).

If Indian animal welfare concerns had prompted Carpenter to establish Cayo Santiago in the 1930s, the polio scare of 1955 renewed US concerns over Indian rhesus supplies. Even as the first effective polio vaccine began to yield results, the polio epidemic of 1955 occasioned a scramble for Indian rhesus macaques, from whose spinal matter the base of the vaccine was extracted. Yet following publicity of the Defense Department's radiation studies, the Indian government placed a moratorium on rhesus exports to the US. India's 1955 agreement with the US on the uses of research primates—prohibiting the use of Indian primates in radiological experiments—sets nationalist statements regarding the value of animals within Indian Prime Minister Jawaharlal Nehru's larger policy of nonalignment. South Asian primatologists played important roles in rallying formerly colonized nations against the use of rhesus in US radiological experimentation. While the Department of Defense used hundreds of Indian monkeys in neutron bomb research and other experiments from the mid-1940s, by the 1970s, several African and Asian countries openly denounced such experiments in international forums and through export bans in place to the present day (Haraway 1989).[6] Even though researchers were able to reestablish access with new assurances that Indian animals would not be used in military research, George Burch emphasized the Indian threat to biosecurity "The problem of supplying monkeys is now acute and there is no assurance whatever that it will improve. The immediate reason is the necessity for poliomyelitis vaccine, but in the recent past an embargo on Indian monkeys was nearly catastrophic" (Burch 1958). In the committee's proposal for the Yerkes Regional Primate Research Center, the need to establish control of the primate supply is even more directly evident:

> In some areas, such as in India, the decimation of the monkey population has led to an appalling reduction in the quality of the animal imported for research. Also, political and religious considerations now loom large, so that the exportation of rhesus monkeys will continue to be curtailed. Embargoes have been placed on rhesus monkeys in the past and it is likely they will be again.... Grave concern was expressed when by 1955 the Indian government had placed two embargoes on the shipment of monkeys. It is also known that the Indian government is assuming a role of responsibility in determining what sorts of research are acceptable for the use of rhesus monkeys before they can be exported to this country. (Burch, n.d.)

As Leon Schmidt noted in a planning meeting for the establishment of the regional primate research centers, importation of rhesus monkeys was seen as a national security priority on par with the raw materials of the US war machine:

> The rhesus monkey is almost as strategic material as tungsten and tin and natural rubber.... We may do to the rhesus what we've almost done to the buffalo here in the United States. It would seem therefore, that any device which would bring together people who could make maximum national use of any of the animals would be a very worthwhile procedure. (CENCPC 1957, p. 19)

Given the troubled history of rhesus importation from India, the committee initially set its sights on other locales and other primate species. Cold War concerns were always on scientists' minds. The 1957 meetings included a significant focus on the decades-old Soviet baboon colony at Sukhumi, Georgia. In the 1950s, Dr. Boris Lapin, who did cardiovascular research on baboons at Sukhumi, met with NIH scientists including Burch and James Watt in Georgia (Garden 1992). Described as both a tourist destination and a home for luminaries such as Pavlov, the US American scientists on the Primate Committee marveled at the ability of the Soviet state to maintain the colony continuously through periods of war. Contrasting this representation with his experience at the underfunded colony in Puerto Rico, Dr. Watt concludes that proper and sustained investment was the key to the Soviet's success.

In assessing other possibilities for primate procurement, central Africa became an obvious choice for the researchers. Two research sites were considered. The first, approved by the outgoing Belgian Congo government in 1959, would establish a new field station for capture and breeding of chimpanzees on a river island in the vicinity

[6]Recently, the National Academy of Sciences has called for an end to Indian restrictions, using the growing presence of HIV in India as a justification (Hearn 2003).

of Stanleyville (present-day Kisangani). The second, the Darjani Primate Center near Kibwezi, Kenya, would house a variety of primate species and was spearheaded by the Southwest Foundation of San Antonio, Texas, a private research group.

Letters from Belgian researchers concerning the Congo station reflect how the dramatic struggles for Congolese independence thwarted US and Belgian officials' attempts to maintain control of chimpanzee habitat. White primatologists living in Congo initially hoped that international support, primarily from US researchers and the UN, would ensure funding after the July 1, 1960, departure of the Belgian government. In the area around the proposed station (called Institut pour la Recherche Scientifique en Afrique Centrale, IRSAC), Kivu leaders apparently initially wanted IRSAC to help maintain the colonial parks that had been retaken by the Congolese. Painting a positive picture for the future in a letter to the NIH requesting temporary financial assistance, IRSAC Director Louis van den Berghe assures the American researchers that "the local Congolese Government of the Kivu was well intentioned, our staff and the population around us were even more friendly than before" (van den Berghe 1960b; Rahm 1960). Yet this cautious optimism perhaps overlooks the fact that the clear racial ordering established in scientific institutions made them targets of Africanization campaigns. At IRSAC, white scientists were highly specialized, highest paid, and last to be fired when institutions closed. The American researchers ultimately placed the project on hold once the Congolese began to expel whites and UN peacekeepers were deployed; they received several letters requesting help in airlifting research chimpanzees out of the country. In an October 1960 plea for help from Karl Meyer to the Director of NIH, James Shannon, Meyer explains that all along, van den Berghe's plan for the US-sponsored chimpanzee colony had been to ensure funding after independence (Meyer 1960).

Van den Berghe himself appears to have understood the US American interest in research chimpanzees to exceed medical science; it was rather one battle in the ongoing Cold War to protect the world from the encroachment of Soviet communism. In an illuminating letter addressed to "an American Friend" (presumably Meyer) 2 months after independence—unsigned perhaps for fear it would be intercepted—van den Berghe gives an extended diagnosis of the colonial failures leading up to IRSAC's precarious situation. Titled "The Congo Situation," the letter sets forth a number of contrasts between Soviet investment in the Congo and lack of US presence. In addition to investing in scholarships, hospitals, and technical institutes in the Congo (van den Berghe calls for a "Tuskegee Institute" for Africa), he notes the Soviets' broad presence across the continent. Emphasizing that the fate of Africa rests on "the fate of the Congo," and that "an effective presence of the USA in all parts of the Congo is essential," van den Berghe ends his four-page letter on the fate of Congolese science and education with a bit of science fiction: "I do not want to make any comments on geopolitics and the actual struggle for power. I have spoken as a scientist and a man devoted to Africa and its inhabitants" (van den Berghe 1960a). Despite the ultimate failure of the Congo project, the Primate Committee saw other prospects in Kenya. With the support of US oil tycoon and amateur cryptozoologist Tom Slick, the US Congress allocated funding for the Southwest Foundation for Research and Education's project at Kibwezi, Kenya. Slick carried out his own expeditions to Nepal and other sites worldwide in search of the Yeti and other zoological legends. He had promised to return the Yeti to the labs of his Southwest Foundation for study (Coleman 1993). The congressional allocation was unusual—it was the only primate funding that was ultimately approved for the maintenance of an institution outside of US territorial control (Animal Resources Program 1968). By 1961, construction and research were in progress on the baboon station in Kibwezi. The situation in an independent Kenya was quite different from that of the Belgian researchers in the Congo. Despite the fact that Kenya was in a similar moment of transition away from colonial rule, the Southwest Foundation was able to secure a lease from the colonial Department of Lands in advance of the 1963 independence declaration. This was a 33-year lease for over 250 acres of unsurveyed land beginning January 1, 1961 (Southwest Foundation for Research and Education, n.d.). At the same time, foundation officials bragged of their decision to pass over more expensive English construction contractors and use an African company. Yet despite this enthusiasm for indigenous contractors, the Southwest Foundation plan was able to allocate widely divergent pay scales for white, Asian, and African workers, with further divisions for skilled and unskilled African laborers (Werthessen 1960, 1961).

Primate studies of the era remained, however, connected to the circuits of exchange underpinning zoological expeditions and the production of sensational colonial discourses around nonhuman animals. According to anthropologist Claud Bramblett, manager of the Southwest Foundation's Darjani station in Kenya from 1963 to 1964, after an American film crew came to the area to film scenes for Howard Hawks' 1962 safari film *Hatari!* starring John Wayne, the Darjani station used the crew's generator to power the station. The Southwest Foundation had initially worked from a safari camp called Bushwackers before being able to fund and construct the Darjani station. During the approximately 5-year tenure of the station, the most expensive local safari expeditions would come to the Darjani area to hunt the large populations of wild mammals including elephants. A pamphlet promoting the station to US American researchers even advertised the possibility of recreational hunting (Southwest Foundation for Research and Education, n.d.). The Darjani station, however, closed in the mid-1960s and returned its lease to the Kenyan government (Bramblett 2008; Vagtborg 1973).

5.4 Turning Inward: Affection and the Domestication of the Macaque

The demise of the African institutions meant that the committee's remaining options were within the existing frameworks of Indian importation and domestic laboratory breeding. This also meant that the rhesus macaque would take center stage as sites for the extraction and reproduction of other species such as chimpanzees were exhausted. In the discussions of the regional primate research centers, there was no consensus on how best to supply and house macaques. Should scientists rely primarily on importation of foreign stocks, where breeding occurs on the free range, but where control over the resource is in the hands of other countries with other interests? If imported, should breeding also occur in free-ranging settings that simulate primates' indigenous environs, as in Cayo Santiago? Or rather than employing these "natural settings," should the state support efforts to undertake more carefully monitored breeding indoors? With disease and genetic variations proving to be the main obstacles to the effective use of rhesus monkeys in research, what institutional organization could provide healthy and increasingly standardized populations of rhesus, either in captivity or in "the wild"? And perhaps most important, would the state be able to quarantine monkeys not just from disease but also from the recurring threats of war, chronic underfunding, and nationalist criticisms of Western biosecurity initiatives?

Debates regarding the proper environment of the experimental macaque include moments in which monkeys are not simply viewed as resources, but also as intentional beings with complex minds and needs. This understanding is promoted by the idea that recognizing the psychological as well as biological needs of animals improves research. In the 1957 meetings of the committee, a major controversy centered on whether keeping monkeys in captivity went against proper animal care. Harry Harlow and Leon Schmidt argued in favor of indoor facilities, speaking of Harlow's success maintaining rhesus in "very small quarters." By contrast, Irving Wright, a Cornell cardiologist, was "not quite sure that locking monkeys up in cages is really giving them a natural environment, if you are going to study their long-term and generation after generation effect on them… I wouldn't expect it to be true of humans; and I don't see why it would be true of monkeys." George Burch added that "it might change their personalities," with Wright implying further that captivity might "affect their outlook on life." Recognizing the labor-value of the rhesus, James Watt, former researcher at Cayo Santiago and director of the National Heart Institute, argued for both captive and semi-free-ranging programs. When "talking about a colony," Watt said, researchers are "shooting for… a diverse group of workers" (CENCPC 1957, p. 25–26). The differences in opinion here betray a logic of animal modeling that contradictorily humanized monkey bodies and minds and objectified their bodies as machines for vivisection.

The NIH program ultimately would establish a mix of indoor and outdoor facilities that could meet a variety of conceptions of best practices. However, it was largely due to the work of University of Wisconsin comparative psychologist Harry Harlow, who had recently demonstrated the efficacy of indoor captive breeding and study, that the rhesus macaque was both broken of its association with tropical nature and effectively humanized, setting the stage for the macaque to become a domesticated figure of national progress. Harlow's laboratory at the University of Wisconsin produced experimental technologies for telling scientific stories about mother-infant love and affection in rhesus that were in turn generalized to humans. His labs were also successful breeding facilities where he developed isolation techniques for ensuring disease-free (specifically TB-free) monkeys that he eventually sold to other researchers in large numbers (Harlow 1986; Haraway 1989). Harlow's breeding techniques—which would eventually include an experimental "rape rack"—were connected to his affection studies, which required the isolation of infant monkeys in order to test their tactile relationships with surrogate mothers that the researcher had fashioned out of wire and cloth (Harlow 1958).

Harlow, who attended the key meeting of the CENCPC, argued for the ability of the experimental scientist to generalize between species in research (Harlow et al. 1972). Harlow's conception of the macaque made this generalizability innate rather than dependent on variables such as the supposed naturalness of the free-ranging habitat. Dismissing the charge of anthropomorphism as an old-fashioned minding of hard-and-fast boundaries between human and animal, Harlow's work made rhesus infants widely publicized models of human love and depression. Images of his rhesus affection experiments were reproduced in textbooks and other visual media. In bringing the rhesus indoors—establishing an ideologically loaded relation of the macaque to the mother, emphasizing its affective life and complexity, and publicizing universal psychological characteristics shared with humans—Harlow worked to establish a new vision of the rhesus divorced from the jungle habitat and expedition narratives that had so sensationally racialized gorillas and chimpanzees and depicted them as hypermasculine monsters. Making the laboratory into a home, the rhesus was rendered domestic—cut off from the transnational circuits of importation, placed under the care of a mother, and removed from the old associations of mimicry and circus performing.

As Harlow's career-making experiments in rhesus affection were informing his advice to the committee on the potential of captive breeding, the laboratory primate as well as the scientist were given makeovers in the US cinema. Echoing the 1953 horror film *Robot Monster*, which idealizes the scientist as "better than a cowboy," films of the early 1950s make the scientist into a model of American manhood and transform the nonhuman primate into his experimental child rather than an overpowering and aggressive monster from the jungle. 1952's *Monkey Business* starring Cary Grant, Ginger Rogers, and Marilyn Monroe and 1952's *Bedtime for Bonzo* starring Ronald Reagan move into a decidedly modern laboratory, celebrating the intentions (if not the abilities) of the white male scientist rather than depicting the pathological possibilities of traversing species boundaries in the lab. From this celebration of American technoscientific modernity, apes not only mirror the human but become science's conduits for envisioning the future of the human. *Bedtime for Bonzo* provides an apt idealization of Harlow's humanization of the laboratory primate, as the chimpanzee is a psychology professor's model for an experiment to prove that the environment of the nuclear family can teach moral values, disproving those who argue that a chimpanzee by nature has no capacity for morality.

If Hollywood producers paid little attention to accurately distinguishing between laboratory monkeys and apes, NIH officials were actively promoting the broader designation "primate" to apply to the entirety of the research that would be carried out in the primate centers. Keeping apes in the institutions, despite the centrality of rhesus subjects, served to maintain the public's association of the project with science rather than older views of mimicry and humor attributed to monkeys. In another 1957 meeting of the committee, Theodore Ruch explicitly addresses the politics of primate representation. He argues for considering including "anthropoid apes" in the agenda for the primate centers primarily because of the "public relations" advantage of not being associated only with monkeys: "There are certain advantages of considering them [apes] in this picture, if for no other reason so that they won't be labeled as the monkeys. It helps to support the word primate. It has a lot better public relations value" (CENCPC 1957). Ruch's statements help us understand the constructedness of the category "primate" in national policy on biomedical research. To advertise "primate research" rather than "macaque research" (or even "chimpanzee research") avoids a range of racialized associations with monkeys and apes, either as sites of danger or humor. Species designations work to attribute value, and the abstraction of "primate research" demonstrates how experimental and institutional language works to shape new associations between animals and research programs.

Part of the appeal of including the ape under the category "primate research" was also its long-time status serving as a "missing link" between human and animal. This association occupied a central place in the colonial anthropological imaginary from the nineteenth century into the early decades of the twentieth. As Donna Haraway argues, however, this association was radically transformed in the aftermath of WWII, as both the Nazi genocides and the global movements for decolonization resulted in the ape bearing the burden of the global history of racism. In this manner, nonhuman primates (particularly ape figures) were simultaneously situated in a temporal difference (figured as the historical past of the human) and universalized as the heritage of all humanity (Haraway 1989). Researchers consistently emphasized the similarity of the primate to the human as justification for the research priority, and this required further moves to rescue the image of the primate. Ruch was not alone in working to address the association of monkeys and apes with primitivity. In the rough draft of George Burch's first proposal for the primate centers, a key word is crossed out in the first sentence: "subhuman." Burch handwrote "nonhuman" in the space above to describe primates in more neutral terms (Burch 1958). The term "subhuman"—used widely by Harlow and many other primate researchers—was present in many of the early documents associated with the founding of the NIH centers, but was eventually replaced in bureaucratic language with "nonhuman."

This properly Darwinian adjective—which today modifies "animal" in the writings of many an animal activist and animal studies scholar—grew in usage out of a set of needs to institutionalize and standardize the primate, which required resignifying it as the proper model for the human. This change in language signals a complex series of transformations in the relation of biomedicine to nonhuman primate bodies. For while it was necessary to domesticate the rhesus and other primates used in research, breaking them of racialized associations with the colonial jungle, this act at the same time allowed for both the humanization of primates and their emergence into politics as imagined rights-bearing subjects once they were no longer "subhuman." Just a year after Burch's editing correction, Jane Goodall would begin her important studies of chimpanzees in Kenya—studies which would be the basis of the next decade's sentimental images of even free-ranging apes as kin of humanity and of field research as a feminized pursuit of a planetary ecological ethic (McHugh 2009–2010). Primate representation was integral to the emergence of a public critique of anthropocentrism: first through new forms of animal activism, later through elaborate visual culture images of ape and monkey emotion, intelligence, and behavior, and most recently through an elaborated academic discourse in "animal studies." Notably, within 10 years of Burch's nominal elevation of primate species on to an equal plane with the human, a new form of animal activism emerged in the US that broke from earlier humane discourse by emphasizing the rights of animals and the duties of humans to incorporate them into political consideration (Rupke 1990; Francione 1996). This act of renaming, then, had the unintended consequence of later helping to identify primate labs as prime targets for animal rights activism. This would solidify the containment of primates as laboratories responded to animal liberation activism through a militarized securitization of laboratory space.

5.5 Conclusions

In her classic sociocultural study of primatology, *Primate Visions*, Donna Haraway documents a number of major transformations in the technologies and ideological underpinnings of primatology as it moved from a minor field of research to a central interdisciplinary site of investigation within US and global scientific discourse. Haraway situates the work of both Carpenter and Harlow within shifting conceptions of communication and embodiment that framed wartime science in the US. Specifically, Haraway associates Carpenter's interest in the social organization and behavior of rhesus within a broader decentering of the human within its social and technological environments, making the command and control of human and animal populations a central scientific and economic interest. Carpenter's later experiments on neurological control of monkeys are the most radical extension of this vision. If for Haraway, Carpenter represented an attempt to establish sovereign power over monkey populations and to capture their supposedly machinelike properties, Harlow's work represented an attempt to grapple with rhesus as emotional creatures. This was paradoxically accessed through the underside of love, the sadism inherent in his experiments that produced the emotional deprivation of the experimental subject in order to prove the existence of its converse, love. While these two primatological figures cannot be simply seen as products of their times, their research indicates a certain shift from a wartime logic of control and command of primates as raw, exploitable, foreign biocapital to a domestication (nationalization/endogamization) of rhesus as humanized beings who could serve a variety of social functions.

Such a shift was influenced as well by the twin geopolitical developments of the US ascendance as superpower and the refusal of scientific access to macaque bodies by newly independent nations in Africa and South Asia. In the process, US American views of rhesus macaques and other laboratory primates shifted, moving the primate out of the jungle and colony and into the laboratory and the home. In the process, the traces of racial and gendered fear of nonhuman primates were often expunged from popular representation, and the category "primate" took on public significance, eclipsing the specificity of macaque or chimpanzee in research institutional promotion. This change had important impacts for the ways in which not just laboratory primates but animals more broadly are represented in scholarly inquiry and political discourse. For while nonhuman primates have rarely been included in the sphere of liberal equality often sought by mainstream animal rights activists, their position as "other animals" (rather than "subhumans") is mediated by the history of biosecurity policies that needed to desensationalize primate figures and to incorporate them into the projects of nationalist technoscience. This shift shows that scientists' attempts to "think with animals," to use the apt phrasing of Lorraine Daston and Gregg Mitman's recent work on anthropomorphism (2005), are an act of knowledge production with specific and often unpredictable political effects. The material and linguistic terms upon which animals model human biosystems reveal both deep assumptions about the extent to which worlds are shared by different species and political and methodological calculi by which researchers attempt to demonstrate biomedical progress (Mitchell 2005).

As a cultural theorist of US American biosecurity, my entry into primatology is concerned with the ways in which the exercise of biosecurity unites particular institutional languages around bodies and species, specific technologies for engineering immunity and other national defenses, and specific geographic imaginaries that frame our cultural conceptions of what national risk and defense might be. If initially, contact with the colonial world was seen as a risk, the colonized bodies of rhesus and chimpanzees were ultimately nationalized and integrated into the very bodies of the inoculated US American public. Rhesus macaques' utility in protecting the nation against disease was initially balanced by a fear over their origins and bodies; and yet a laboratory science that could domesticate these creatures—turn them from exotic objects into tamed national subjects—was an important component in the transformation of US society into one that began to accept increasingly radical biomedical intervention into patient bodies at the same time that it came to see humans as closely related to other primates. The conscription of rhesus as defenders of the nation against dread diseases, then, demonstrates that cooperation and conflict between humans and rhesus are conjoined and are substantially mediated by the language, visual representations, and institutional and capital circuits through which we forge our relationships to these other animals.

References

Anderson W (2006) Colonial pathologies: American tropical medicine, race, and hygiene in the Philippines. Duke University Press, Durham/London

Animal Resources Branch, Division of Research Resources, NIH (1963) Animal resources for biomedical research. George E. Burch Papers. National Library of Medicine, Bethesda

Animal Resources Program, Division of Research Resources, NIH (1968) Regional primate research centers—the creation of a program. Public Health Service, Bethesda

Anonymous (1939) First American monkey colony starts on Puerto Rico Islet. Life 2:26

Bederman G (1993) Manliness and civilization: a cultural history of gender and race in the United States, 1880–1917. University of Chicago Press, Chicago/London

Bramblett C (2008) [Personal Communication] 20 Aug

Brennan T (2005) The economic image-function of the periphery. In: Loomba A, Kaul S, Bunzl M, Burton A, Esty J (eds) Postcolonial studies and beyond. Duke University Press, Durham/London, pp 101–122

Burch GE (1957) A proposal for a national primate research station. Unpublished draft, Box 43, "National Advisory Heart Council-Primate Colony, 1957–8", George E. Burch Papers. National Library of Medicine, Bethesda

Burch GE (n.d.) A program for the Yerkes laboratories. Unpublished draft, George E. Burch Papers. National Library of Medicine, Bethesda

Carpenter CR (1940) Rhesus monkeys (*Macaca mulatta*) for American laboratories. Science 92:284–286

CENCPC (1957) Meeting minutes, 25 Sep. Unpublished transcript, Box 43, "National Advisory Heart Council-Primate Colony, 1957–8", George E. Burch Papers. National Library of Medicine, Bethesda

Clarke A (1998) Disciplining reproduction: modernity, American life sciences, and "the problem of sex. University of California Press, Berkeley

Clause BT (1993) The Wistar rat as a right choice: establishing mammalian standards and the ideal of a standardized mammal. J Hist Biol 26:329–349

Coleman L (1993) Tom slick and the search for the yeti. Faber and Faber, Boston/London

Cook-Deegan R, McGeary M (2006) The jewel in the federal crown? History, politics, and the National Institutes of Health. In: Stevens R, Rosenberg C, Burns L (eds) History and health policy in the United States. Rutgers University Press, New Brunswick, pp 176–201

Daston L, Mitman G (eds) (2005) Thinking with animals: new perspectives on anthropomorphism. Columbia University Press, New York

Duany J (2002) The Puerto Rican nation on the move. University of North Carolina Press, Chapel Hill/London

Foucault M (1995) Discipline and punish. Vintage, New York

Foucault M (2003) "Society must be defended": lectures at the collège de France, 1975–76. Picador, New York

Francione G (1996) Rain without thunder: the ideology of the animal rights movement. Temple University Press, Philadelphia

Garden V (1992) Interview with William Gay, 15 July. Downloaded from http://history.nih.gov/NIHInOwnWords/docs/gay_01.html

Haraway D (1989) Primate visions: gender, race, and nature in the world of modern science. Routledge, London/New York

Harlow HF (1958) The nature of love. Am Psychol 13:673–685

Harlow CM (1986) The evolution of Harlow research. In: Harlow CM (ed) From learning to love: the selected papers of H. F. Harlow. Praeger, New York, pp xxxiv–xxxv

Harlow HF, Gluck JP, Suomi SJ (1972) Generalization of behavioral data between nonhuman and human animals. Am Psychol 27:709–716

Hearn J (2003) Primate priorities – an international perspective. In: Institute for laboratory animal research, international perspectives: the future of nonhuman primate resources. Proceedings of the workshop held on 17–19 Apr 2002. National Academies Press, Washington, pp 1–10

Kaplan A (1993) "Left alone with America": the absence of empire in the study of American culture. In: Kaplan A, Pease D (eds) The cultures of United States imperialism. Duke University Press, Durham/London, pp 3–21

Kohler R (1994) Lords of the fly: Drosophila genetics and the experimental life. University of Chicago Press, Chicago/London

Lakoff A (2010) Two regimes of global health. Humanity 1:59–79
Latour B (1993) The pasteurization of France. Harvard University Press, Cambridge/London
Locke CM (1938) Peopling an island with gibbon monkeys: an ambitious West Indian experiment in biology. The Illustrated London News, 13 Aug
Martin E (1994) Flexible bodies: tracking immunity in American culture—from the days of polio to the age of AIDS. Beacon, Boston
May ET (1990) Homeward bound: American families in the cold war era. Basic, New York
McHugh S (2009–2010) Sweet Jane. Minn Rev 73–74:189–203
Merkel H (1997) Quarantine! East European Jewish immigrants and the New York city epidemics of 1892. Johns Hopkins University Press, Baltimore
Meyer K (1960) [Letter to James Shannon] 3 Oct. Box 43, "NIH-Primate Research Centers-Committee on Organization, July 1960-June 1961", George E. Burch Papers. National Library of Medicine, Bethesda
Mitchell SD (2005) Anthropomorphism and cross-species modeling. In: Daston L, Mitman G (eds) Thinking with animals: new perspectives on anthropomorphism. Columbia University Press, New York, pp 100–118
Nadel A (1995) Containment culture: American narratives, postmodernism, and the atomic age. Duke University Press, Durham/London
Noble DL (2006) Operations in another time: a U.S. naval intelligence mission to China in the 1930s. Studies in intelligence 50 (unclassified edition). Downloaded from https://www.cia.gov/csi/studies/vol50no2/html_files/Mission_China_3.htm
Paul JR (1971) A history of poliomyelitis. Yale University Press, New Haven/London
Rahm U (1960) [Translated letter to Karl Meyer] 24 Oct. Box 45, "Primate Research Centers Belgian Congo Trip, 1960-1," George E. Burch Papers. National Library of Medicine, Bethesda
Rawlins RG, Kessler MJ (1986) The history of the Cayo Santiago colony. In: Rawlins RG, Kessler MJ (eds) The Cayo Santiago macaques: history, behavior and biology. State University of New York Press, Albany, pp 13–46
Rogers N (1992) Dirt and disease: polio before FDR. Rutgers University Press, New Brunswick
Rony FT (1996) The third eye: race, cinema, and ethnographic spectacle. Duke University Press, Durham/London
Rose N (2007) The politics of life itself: biomedicine, power and subjectivity in the twenty-first century. Princeton University Press, Princeton/Oxford
Rupke NA (1990) Vivisection in historical perspective. Routledge, New York/London
Russell N (2002) The wild side of animal domestication. Soc Anim 10:285–302
Serlin D (2004) Replaceable you: engineering the body in postwar America. University of Chicago Press, Chicago/London
Shah N (2001) Contagious divides: epidemics and race in San Francisco's Chinatown. University of California Press, Berkeley
Shell M (2005) Polio and its aftermath. Harvard University Press, Cambridge/London
Shohat E, Stam R (1994) Unthinking eurocentrism. Routledge, New York/London
Shukin N (2009) Animal capital: rendering life in biopolitical times. University of Minnesota Press, Minneapolis/London
Southwest Foundation for Research and Education (n.d.) [Unpublished pamphlet on Darjani]
Stern AM (2005) Eugenic nation. University of California Press, Berkeley
Stoler AL (1995) Race and the education of desire: Foucault's history of sexuality and the colonial order of things. Duke University Press, Durham/London
Vagtborg H (1973) The story of the southwest research center: a private, nonprofit, scientific research adventure. University of Texas Press, Austin/London
Van den Berghe L (1960a) [Letter to Karl Meyer] 29 Aug. Box 43, "NIH-Primate Research Centers-Committee on Organization, Jan-May 1960", George E. Burch Papers. National Library of Medicine, Bethesda
Van den Berghe L (1960b) [Letter to Karl Meyer] 17 Sep. Box 43, "NIH-Primate Research Centers-Committee on Organization, July 1960-June 1961", George E. Burch Papers. National Library of Medicine, Bethesda

Wadiwel D (2002) Cows and sovereignty: biopower and animal life. Borderlands 1. Downloaded from http://www.borderlands.net.au/vol1no2_2002/wadiwel_cows.html

Wald P (2008) Contagious: cultures, carriers, and the outbreak narrative. Duke University Press, Durham/London

Werthessen NT (1960) [Letter to Hal Eyestone] 18 June. Box 44, "Primate Research Centers-Correspondence, 1958-60", George E. Burch Papers. National Library of Medicine, Bethesda

Werthessen NT (1961) [Letter to Drs. Meyer, Burch, Stare, and Eyestone] 30 Jan. Box 44, "NIH-Primate Research Centers-Project Site Visits, 1960–1", George E. Burch Papers. National Library of Medicine, Bethesda

Wilson D (1998) A crippling fear: experiencing polio in the era of FDR. Bull Hist Med 72: 464–495

Wolfe C (2010) Before the law: animals in a biopolitical context. Law Cult Humanit 6:8–23

Chapter 6
Macaque Tourism: Implications for Their Management and Conservation

Avanti Mallapur

6.1 Introduction

Humans have lived in close proximity to non-human primates for centuries (Hill 1999). Yet today, burgeoning human numbers and fragmented forest habitats threaten the continuance of this association, raising serious concerns among scientists and conservationists about our ability to share ecological space with non-human primates. Macaques live in diverse habitats from Japan to Afghanistan in Asia and in Gibraltar and Morocco in Northern Africa. With their ability to colonize degraded landscapes, they possess far-reaching ecological and conservation significance, particularly in their roles as potential pollinators and seed dispersers in human-modified habitats (Lucas and Corlett 1998; Nakashima et al. 2008; Tsujino and Yumoto 2009). Hence, it is imperative that macaques are protected from factors that could reduce their population numbers in the wild and cause their extinction. Plummeting numbers have already caused several macaque species to be listed as near threatened (e.g. *Macaca assamensis* and *M. thibetana*), endangered (e.g. *M. maura, M. silenus, M. sylvanus, M. sinica and M. munzala)* or critically endangered (e.g. *M. pagensis and M. nigra*), reiterating the urgent need to conserve these species and protect their habitats (IUCN 2010).

In recent years, humans' interest in foreign landscapes, cultures, food and people have grown (Gössling 2002; Hill 1999), including wildlife-protected areas where they can see and interact with animals (Orams 2002). Primate-inhabited protected areas are popular tourist destinations, where tourists sometimes feed and interact with free-ranging non-human primates, and primate ecotourism is now considered a new and improved pathway to introduce the natural environment to local communities (Fuentes 2006; Hsu et al. 2009; McCarthy et al. 2009; Chauhan and Pirta 2010).

A. Mallapur (✉)
Director of Biodiversity Conservation, Department of Biodiversity Conservation,
Aaranyaa, 1029 Ray Andra Drive, DeSoto, TX 75115-3915, USA
e-mail: avanti.mallapur@aaranyaa.org

S. Radhakrishna et al. (eds.), *The Macaque Connection: Cooperation and Conflict between Humans and Macaques*, Developments in Primatology: Progress and Prospects 43,
DOI 10.1007/978-1-4614-3967-7_6, © Springer Science+Business Media, LLC 2013

Several field conservation programmes for macaques have recorded them to be popular tourist attractions. However, it is a multifaceted approach and could introduce free-ranging macaque populations to new pressures that have not been studied or documented. For example, food provisioning by tourists has in some cases resulted in aggressive interactions between macaques and humans (Fuentes 2006; Hsu et al. 2009; McCarthy et al. 2009; Chauhan and Pirta 2010), and these aggressive responses from macaques negatively influenced the attitudes of humans towards them (Sha et al. 2009). Close interactions such as these also increase the likelihood of anthropozoonotic disease transmission between the two species (Jones-Engel et al. 2001; Goldberg et al. 2007; Muehlenbein et al. 2010). With burgeoning human populations, an increasing overlap between macaque habitats and human dwellings has been recorded in countries such as India, Indonesia and Gibraltar and scientists have observed higher levels of competition between macaques and humans in regions where their habitats overlap. Macaque behaviour has been found to notably change in such situations, with higher incidences of aggressive encounters with young, resulting in infant mortality. Macaques living in and near areas that are frequently visited by humans such as temples tend to interact more with them and be regularly fed, especially in countries where primates are worshipped. In zoos and wild animal parks, captive macaques are also considered popular attractions. Large numbers of zoo tourists have been observed to have a detrimental effect on captive macaque behaviour and welfare, influencing displays of behavioural abnormalities. It is imperative that programmes that are interested in utilizing tools such as ecotourism for species conservation are well planned to reduce the probable threat to free-ranging macaque populations and their natural habitats. In this chapter, I propose to explore the multiple impacts of tourism in macaque habitats and its influence on free-ranging and captive macaque populations.

6.2 What Is Ecotourism?

Humans visit protected macaque-inhabited areas (Fuentes 2006; Ram et al. 2003; Hsu et al. 2009; McCarthy et al. 2009) for recreational purposes or tourism. The International Conference on Travel and Tourism Statistics in Ottawa, Canada, defined a tourist as a person "travelling to and staying in places outside their usual environment for not more than one consecutive year for leisure, business and other purposes not related to the exercise of an activity remunerated from within the place visited" (WTO 1995). Tourism is a popular international recreational interest with approximately 940 million tourist arrivals in 2010 (UNTWO Tourism Highlights 2011). Developing countries therefore regard tourism as one of their important economic sectors (UNWTO Annual Report 2010: 11).

In the 1980s, environmentalists introduced the concept of ecotourism – an action that would provide tourists visiting nature for recreation, an understanding of the natural habitat in order to instil an appreciation and respect for that environment (Honey 1999). Ecotourism was defined as "responsible travel to natural areas which conserves the environment and improves the welfare of local people" (Western 1993). Well-established ecotourism practices monitor natural habitats frequented by tourists and minimize the harmful effects of conventional tourism on the environment while at the same time providing economic support to local communities (Randall 1987).

In many Asian countries, wildlife-protected areas cater to both tourists and ecotourists and, thus, the terms are used interchangeably. Those visitors who request a tour with a nature guide and show an interest in knowing more about the wildlife-protected areas and its conservation issues and in some cases live in a forest lodge or ecolodge to experience the environment are referred to as ecotourists. Many wildlife-protected areas have temples and other religious institutions located within them. Travellers that enter these protected areas in order to visit the temples are considered tourists. Several wildlife-protected areas also have transport lines such as roads and railway tracks running through their forests. Travellers driving through these forests are also regarded as tourists. However, it is challenging to differentiate "tourists" from "ecotourists" especially when both visit the same protected area. Several papers cited in the text below refer to travellers as tourists instead of ecotourists and vice versa. In order to address this issue and examine the interactions of travellers with wild macaques and their influence on the environment of wildlife-protected areas, I refer to all travellers as tourists in the following text.

6.3 Macaques, Tourists and Provisioning

Several research studies, across Asia, have documented that it is common practice for humans to feed macaques. Food provisioning has been documented in Formosan macaques in Taiwan (Hsu et al. 2009), Barbary macaques in Gibraltar (Fuentes 2006), rhesus macaques both in Nepal (Pedersen 1997) and India (Chauhan and Pirta 2010), long-tailed macaques in Indonesia (Fuentes and Gamerl 2005; Fuentes 2006) and Singapore (Fuentes et al. 2008), Tibetan macaques in China (Zhao 1999; McCarthy et al. 2009; Ruesto et al. 2010), bonnet macaques in India (Ram et al. 2003) and Assamese macaques in Nepal (Regmi and Kandel 1998).

Food provisioning habituates macaque populations to human presence, and studies have shown increased levels of aggressive behaviours by macaques towards tourists in such contexts (O'Leary and Fa 1993; Hill 1999; Fuentes and Gamerl 2005; Fuentes 2006; Hsu et al. 2009; McCarthy et al. 2009; Chauhan and Pirta 2010). A study on interactions between Formosan macaques and tourists at a nature reserve in Taiwan (Hsu et al. 2009) found food provisioning to significantly increase durations and frequencies of aggressive interactions between macaques and tourists. Most of the observed interactions were tourist-initiated (approximately 71%). Interestingly, adult male macaques and tourists were considered most likely to engage in agonistic interactions and accounted for 16.4% of overall interactions. Aggressive behaviours directed by macaques during food provisioning (60.2 times/s) were 4.87 times greater than during non-food provisioning (12.36 times/s).

The link between agonistic responses of macaques and food provisioning by tourists has also been observed by other scientists (Wheatley and Harya Putra 1994; Fuentes and Gamerl 2005; Fuentes 2006; Chauhan and Pirta 2010). In the Khao Yai National Park, staff and scientists recorded behavioural changes in wildlife due to feeding by tourists (Phumsathan and Nepal 2008). Most of the food provisioning was done along the roadside within the park that led to wild macaques spending 83.3% of their time by the road in anticipation of being fed. Free-ranging macaques were found to tailor their diurnal activity patterns to tourist visitation schedules to maximize the probability on being fed (McCarthy et al. 2009). These individuals were also less active than the unhabituated macaques and spent twice as much time close to tourists, sometimes feeding on five times as much anthropogenic food (Patzschke et al. 2000). Again, food provisioning by tourists to free-ranging bonnet macaques in Mudumalai Wildlife Sanctuary, southern India, noticeably increased intraspecific food competition among females (Ram et al. 2003).

Clearly, food provisioning has serious impacts on the behaviour and social interactions of macaque individuals. And in this context, it is important to examine why humans feed macaques. Wildlife feeding is used as a means by which tourists and their guides create opportunities to approach wild animals in order to observe and interact with them (Orams 2002; Fuentes and Gamerl 2005). They believe that by doing this, they become closer to wildlife and nature. It has been suggested that television programmes such as those on "National Geographic" or "Discovery" and environmental magazines have created a new "awareness", encouraging large numbers of people to travel and "reconnect" with nature (Orams 2002). Diminishing opportunities to interact with the natural environment locally builds a need to nature travel. Another important factor is that of religious beliefs and cultural practices. Animal deities are an important part of the culture in many Asian countries. In Hinduism, often called the oldest living religion (Morgan 1987), for example, it is believed that animals are manifestations of God and that both humans and animals experience reincarnation or life after death (Flood 1996). Humans can be reincarnated as animals and vice versa, which signifies that all forms of life are to be respected. Animals that play an important role either as manifestations of gods or as their vehicles (*vahanas*) are greatly revered in Hindu mythology. The most prominent example of course is Hanuman, the monkey god, worshipped for his power, strength and protection (Lutgendorf 1994). Other Asian cultures and religions also promote respect for or worship wild animals. Both Jainism and Buddhism preach "*ahimsa*" or non-violence towards all living beings (Gopalan 1973). The Ainu people from Japan believe that the bear or "*kamui*" is the head of the gods, and worship him (Kindaichi 1949). Coastal villagers in Vietnam believe that whales bring good luck and prosperity to fisherfolk and thus revere them (Taylor 2003). Some local societies in Sumatra make elephant figurines in stone and wood thus displaying the animal's importance in their communities (Schnitger 1938). People in Hindu societies feed animals in order to obtain good *karma*. Many studies have observed that macaques found around Hindu temples are venerated and provisioned (Ali and Pai 2001; Fuentes and Gamerl 2005; Fuentes 2006; Fuentes et al. 2008), although they may be reviled elsewhere.

6.4 Tourist Behaviour and Its Effect on Macaques

Apart from food provisioning, aggressive responses by free-ranging macaques are also influenced by tourist behaviour (McCarthy et al. 2009; Ruesto et al. 2010). A study conducted by McCarthy et al. (2009) on Tibetan macaques and tourists at Mt. Huangshan, China, showed that 81.9% of macaque–human interactions were initiated by tourists. Some of the behaviours displayed by tourists such as pointing at macaques were found to initiate threats from the species. Similarly, Fuentes (2006) in his research on Barbary macaques and tourists in Gibraltar found significantly high levels of contact interactions between the two species. He also noted food provisioning by tourists, taxi drivers and tour guides. In another study in Gibraltar, O'Leary and Fa (1993) found a similar increased proportion of interactions initiated by tourists (approximately 76%).

Ruesto et al. (2010) examined the impact tourists have on provisioned, habituated Tibetan macaques. Data in the form of tourist densities, behaviours and sound decibel levels were collected from a tourist-viewing platform on Mt. Huangshan in China. Total tourists' behaviours and noise decibel levels positively correlated with threats from macaques. Another study at the same location recorded the tendency of tourists to initiate and continue interactions with macaques until they received a response that was, more often than not, a threat (McCarthy et al. 2009). It was found that threat responses occurred towards tourist behaviours such as point and slap rail. Berman et al. (2007) recorded a similar threat response from macaques to the tourist behaviour ground slap that closely resembled slap rail. The elevated level of noise from these behaviours probably initiated a threat response from the macaques.

Fuentes and Gamerl (2005), in their research on interactions between long-tailed macaques and tourists at the Padangtegal Wanara Wana Temple in Bali, Indonesia, observed 420 aggressive interactions between the two species. Tourists provided the macaques with food during 308 (73%) of the 420 interactions. Adult and subadult male macaques accounted for a majority of the contact aggression (including 48 biting instances) and lunging and chasing behaviours directed towards humans. Contact aggression accounted for 78% of the interactions. A higher proportion of contact aggression (e.g. biting) occurred in Bali when compared to Barbary macaque–tourist interactions in Gibraltar possibly because temple tourists in Bali used the forest frequently for religious ceremonies, leading to increased encounter rates between macaques and tourists there (Fuentes 2006). Temple tourists have other deleterious effects on wildlife-protected areas. Ali and Pai (2001) report that in the Kalakad-Mundanthurai Tiger Reserve in southern India, temple tourists bring radios into the protected area increasing decibel levels in the reserve. Cooking fires and waste litter are other consequences from their visits. During the largest festival in the area, *Adi Amavasi*, approximately 500,000 temple tourists visit the reserve over a period of 10 days.

In their study on the impacts of tourist activities in Thailand's Khao Yai National Park (KYNP), Phumsathan and Nepal (2008) found that tourist presence and behaviour had an impact on the soil, water and vegetation in the park. The major activities in the park were camping and hiking, and these were reported to cause soil erosion, humus layer removal and reduction in organic matter, resulting in the ground becoming bare. The loss of humus and organic matter affected growth of natural vegetation. Several exotic species of shrubs and trees were planted around the tourist centre, and several trails and paths were created for tourists, which affected vegetation composition and diversity. Water quality in the park was also affected. KYNP offers several water-based tourist attractions within its boundaries, which bring tourists in large numbers. Tourists were recorded using toiletries such as soap and shampoo and disposing kitchen waste in the water, which could impact aquatic wildlife.

Free-ranging macaques living in urban environments are also known to be affected by human activities. Chauhan and Pirta (2010) observed that in Shimla, northern India, rhesus macaques spent a considerable proportion of their time obtaining food from humans. Most often (68% of the time), they drank water from the city drain, which could adversely influence their health and welfare. Long-tailed macaques in Singapore are commonly fed by people, often from automobiles. Macaques were also recorded to be more likely to interact with humans if food was present (Fuentes et al. 2008).

6.5 Anthropozoonotic Disease Transmission Between Macaques and Tourists

Anthropozoonotic diseases have been transmitted from free-ranging non-human primate populations to humans (Holmes 1996; Wolfe et al. 1998; 2004; Engel et al. 2002; Jones-Engel et al. 2005; Gillespie 2006). Macaques share a wide range of anthropozoonotic diseases with us; hence, close encounters between human and macaque populations increase disease transmission risks and could potentially lead to poor health and mortality among free-ranging primate populations (Jones-Engel et al. 2001). A significant number of tourists in protected areas could be unhealthy and potentially infectious, creating an unnecessary risk of pathogen transmission to wildlife (Muehlenbein et al. 2010). Apart from tourists, macaques are also at the risk of contracting diseases from observers, tour guides and researchers who spend long durations of time in close proximity with them. Even wildlife rescue and rehabilitation efforts focused on reintroducing and repopulating free-ranging populations increase human–animal contact, thus increasing the risk of disease transmissions. Research stations, where waste is improperly disposed or where the facilities are unkempt and unhygienic, may also cause disease outbreaks that could be transmitted to wildlife populations (Wallis and Lee 1999). It is imperative that observers, researchers and park personnel follow strict biosecurity protocols and regulations to manage and maintain disease-free and hygienic environments. In order to reduce disease transmission risks between tourists and macaques, interspecific interactions should be managed, human behaviours towards free-ranging animals modified and the use of preventive medicine where appropriate and, especially for humans, should be encouraged.

6.6 Captive Primates and Zoo Tourists

The primary goals of modern zoos are (a) to educate visiting public about conserving wildlife, (b) to support in situ and ex situ conservation efforts, (c) to improve the health and welfare of captive wildlife, (d) to encourage zoo-based and field research and (e) to entertain visiting public. Over 100 million people visit American zoos each year (Koontz 1997; Hanson 2002), and a majority visit them for entertainment (Reade and Waran 1996) and to interact with wildlife (Wood 1998).

Approximately 15 zoos in North America house captive macaque populations. Macaques are highly social species and found in a wide variety of natural habitats in a range of climates. They are also known to be good tree climbers. Zoos find it challenging to mimic their natural environments by providing them with an access to the vertical dimension and an optimal social group. Behavioural observation has commonly been used as a non-intrusive research tool to assess the welfare of captive primates (Mench and Mason 2000; e.g. Mallapur et al. 2005; Marriner and Drickamer 1994; Mootnick and Baker 1994; Little and Sommer 2002). Studies have shown that macaque species are sensitive to stressful environments and have been documented to display behavioural abnormalities (Mallapur and Choudhury 2003; Mallapur et al. 2005). A study on captive non-human primates housed in Indian zoos showed that omnivorous primates such as macaques and gibbons displayed some types of abnormal behaviours (stereotypic pacing, floating limb and self-clasping) while folivores (e.g. langurs) did not (Mallapur and Choudhury 2003).

Tourists visiting zoos have been found to be a source of stress to captive macaques (Mallapur et al. 2005). Thirty captive lion-tailed macaques housed in eight zoos across India were observed on "tourist presence" days and on "tourist absence" days, which were zoo holidays. Individuals displayed increased levels of begging on days that tourists were present. Frequencies of abnormal behaviour, self-biting, begging and bouncing were also significantly higher in the presence of tourists. Social behaviours, especially reproductive behaviours, were displayed more on days when tourists were absent. Even the use of enclosure space was affected by tourist presence. Captive lion-tailed macaques were found to spend more time at the back of the enclosure (the furthermost point from tourist-viewing area) when tourists were present. Mallapur et al. (2005) also observed seven captive lion-tailed macaques in on-exhibit and off-exhibit enclosures. There was a marked increase in abnormal behaviours and stereotypic pacing when the captive macaques were housed "on-exhibit" than when they were housed "off-exhibit". Similar to people visiting wildlife-protected areas, zoo tourists are also interested in interacting with wildlife and have a tendency to feed captive primates (Wood 1998). Displays of begging behaviour by captive primates demonstrate their anticipation of being fed. This was clearly documented in the behavioural study conducted on captive lion-tailed macaques in southern Indian zoos (Mallapur et al. 2005). Cooke and Hosey (1995) documented begging behaviour in captive chimpanzees as a response to an interactive sequence of behaviours with zoo tourists and that they were fed 25% of the time when such a behavioural sequence was initiated by zoo tourists. This supports the interpretation that, like free-ranging primates, captive individuals interact with humans primarily to access food.

6.7 Tourism, Conservation Awareness and Macaque Conservation

Tourists typically visit macaque-inhabited habitats with the hope of interacting with these non-human primates. However, close interactions between humans and macaques are detrimental for both species (Wheatley and Harya Putra 1994; Fuentes and Gamerl 2005; Fuentes 2006; Hsu et al. 2009; McCarthy et al. 2009; Chauhan and Pirta 2010). One way to remedy this situation is to initiate conservation awareness programmes that create a learning experience for the visiting public, bearing in mind that they achieve the goals of their visit. Conservation awareness programmes with user-friendly policies for local communities have also been recommended for biodiversity-rich areas that harbour threatened wildlife species (Mishra et al. 2006). Assessment and questionnaire surveys help understand how local communities and tourists interact with nature. Wildlife surveys also provide insights into site-species issues, such as tourist numbers and behaviour, and food provisioning practices. Educating local communities about the significance of conserving native wildlife is essential to improve and sustain species richness in well-traversed national and state forest parks. Safety regulations should also be designed for ecotourism projects that take into account macaque behaviour and thereby reduce the possibilities of macaque attacks. In some regions, conservation education programmes have managed to be successful in promoting primate conservation and habitat protection. For example, workshops have also been used to assemble primatologists, scientists, people working for zoos, local and international government agencies and non-profit organizations, in order to develop guidelines to conserve an endangered hylobatid, the Javan gibbon, *Hylobates moloch* (Supriatna 2006). However, there are limited number of ongoing conservation awareness programmes for macaques in South and Southeast Asia, a region that is inhabited by 21 of the 22 macaque species. The high incidence of agonistic behaviours between macaques and tourists (O'Leary and Fa 1993; Hill 1999; Fuentes and Gamerl 2005; Fuentes 2006; McCarthy et al. 2009; Hsu et al. 2009; Chauhan and Pirta 2010), anthropozoonotic disease transmission (Holmes 1996; Wolfe et al. 1998, 2004; Jones-Engel et al. 2001, 2005; Engel et al. 2002; Gillespie 2006; Goldberg et al. 2007; Muehlenbein et al. 2010) and pet trade (Ahmed 2001; Geissmann et al. 2006; Van Lavieren and Wich 2009; Shepherd 2010) strongly support and reiterate the urgent need to initiate and maintain conservation awareness programmes for tourists and local communities and to share information on the conservation significance of protecting macaques and their habitats. Involving local communities in primate rescue and rehabilitation and in managing nurseries of native plant life to support reforestation of natural ecosystems would be an ideal means of sustaining local economies and conserving primates in their habitats. Sha et al. (2009) conducted a questionnaire survey of tourists and local inhabitants in six zones in Singapore regarding their interactions with long-tailed macaques. The results of the study showed that 33% of the respondents had problematic interactions with the macaques. Tourists faced more problems with the macaques than did local communities, and this affected their behaviour towards the macaques. The survey documented that approximately 88% of the respondents believed that conserving macaques was crucial. The majority of the respondents (67%) were interested in supporting conservation awareness programmes, which could resolve or reduce the macaque–human conflict issues, and only a small group (approximately 15%) supported translocating the macaque populations to other areas. Such data provides an excellent base to build conservation awareness programmes for local communities and tourists.

Zoological parks are well known for their conservation awareness programmes, and being popular tourist destinations are ideal places to educate the visiting public, to raise funds to sustain breeding programmes and to support field conservation projects. American zoos are visited by a large number of people, providing an excellent potential to influence public perception and enhance their knowledge and understanding of wildlife (Caras 1995). In the Fort Wayne Children's Zoo in Indiana, for example, an interactive education and conservation action programme was developed to encourage students to play an active role in conserving wildlife. These students, through their courses, raised funds to support primate conservation in tropical countries like Indonesia (Wiese and Hutchins 1997). Although structured talks for the visiting public form a considerable portion of zoo education programmes, some zoo educationists believe that much of the learning potential that the zoo has to offer its tourists rests in viewing the animals and tourist response to them (Moss and Esson 2010). Several zoos have also initiated conservation awareness programmes in the habitat range countries of the species of their focus, for example, Jersey Zoo in the United Kingdom and National Zoo in the United States. These conservation education programmes are considered crucial for the success of reintroduction projects and other in situ conservation efforts (Kleiman et al. 1986; Durrell and Mallinson 1987).

Although south and Southeast Asia have fragile tropical ecosystems, few zoos support in situ conservation or have education programmes. These ecosystems are prime habitats for free-ranging macaque populations, and zoos in this region could benefit from zoo education programmes. Furthermore, several zoos in this region house macaques in species-specific social organizations providing an ideal environment for awareness programmes on primate conservation. By initiating long-term education programmes in wildlife-protected areas and zoos, funds can be raised to support field conservation efforts, poaching and vandalism by humans towards wildlife and their habitats can be effectively reduced, and local students and adults can be educated on the significance of protecting and monitoring natural environments. To be more effective, staff conducting these education programmes should be able to contribute information on conserving wildlife in both theory and practice (Jacobson 2010). The educators will need to have experience in the fields of planning, collaboration, evaluation and entertainment to make conservation education programmes more successful.

6.8 Tourism and Free-Ranging Macaque Populations in Asia in the Future

Some Asian countries do depend on wildlife tourism for revenue. When the success of the oil industry in Indonesia diminished, the country targeted ecotourism, inviting international tourists to view their wildlife habitats and endemic species such as the Komodo dragon *Varanus komodoensis* and Celebes black-crested macaque *Macaca nigra* (Ross and Wall 2001). However, macaque tourism may have several undesirable consequences in the form of food provisioning, disturbances to macaques and the risk of disease transmission that could prove detrimental to both macaques and humans alike. Care should be taken to be sure to maintain a maximum flight distance between tourists and free-ranging macaque populations and to prohibit food provisioning. Provisioning macaques with human foods can be unhealthy and has led to closer interactions between macaques and humans. Avoiding food provisioning will decrease close interactions between macaques and humans, which in turn will reduce the opportunities for zoonotic disease transmission between the two species.

Conservation awareness programmes should be initiated to increase the potential of raising funds for macaque conservation and for training tourists on the protocols to follow while visiting free-ranging non-human primate populations. Enforcing wildlife and animal welfare laws by encouraging tourists to follow them and fining those who do not should be supported. Several field conservation projects have initiated successful awareness programmes to protect free-ranging primates (Supriatna 2006; Savage et al. 2010; Strier 2010), some leading to reduced use of forest products by human settlements living around non-human primate habitats. Conservation awareness programmes such as these would help us exchange our knowledge with tourists, local communities and the general public with the hope of increasing human understanding and respect for macaques and improving the prospect of saving these species from extinction.

Acknowledgements I would like to thank the editors of this book for giving me the opportunity to write this chapter and the reviewers of this chapter for their invaluable comments. I am indebted to Kiran Mallapur (Ayi) for reading through my drafts and revising them. I would also like to thank Sindhu Radhakrishna, Anand Ramanathan and Divya B U for their advice and comments on my manuscript.

References

Ahmed A (2001) Illegal trade, and utilization of primates in India. Envis Bull Wildl Prot Areas 1:177–184

Ali R, Pai A (2001) Human use areas in the Kalakad–Mundanthurai Tiger Reserve. Curr Sci 80:448–452

Berman CM, Li JH, Ogawa H, Ionica C, Yin HB (2007) Primate tourism, range restriction, and infant risk among *Macaca thibetana* at Mt. Huangshan, China. Int J Primatol 28:1123–1141

Caras R (1995) Zoos and the public A view from the ASPCA In: Norton BG, Hutchins M, Stevens EF, Maple TL (eds) Ethics on the Ark. Smithsonian Institution Press: Washington DC, USA, pp 297–300

Chauhan A, Pirta RS (2010) Socio-ecology of two species of non-human primates, rhesus monkey (*Macaca mulatta*) and Hanuman langur (*Semnopithecus entellus*), in Shimla, Himachal Pradesh. J Hum Ecol 30:171–177

Cook S, Hosey GR (1995) Interaction sequences between chimpanzees and human visitors at the zoo. Zoo Biol 14:431–440

Durrell G, Mallinson J (1987) Reintroduction as a political and educational tool for conservation. Dodo (J Durrell Wildl Conserv Trust) 24:6–19

Engel GA, Jones-Engel L, Schillaci MA, Suaryana KG, Putra A, Fuentes A, Henkel R (2002) Human exposure to herpesvirus B-seropositive macaques, Bali, Indonesia. Emerg Infect Dis 8:789–795

Flood GD (1996) An introduction to Hinduism. Cambridge University Press, Cambridge

Fuentes A (2006) Human culture and monkey behavior: assessing the contexts of potential pathogen transmission between macaques and humans. Am J Primatol 68:880–896

Fuentes A, Gamerl S (2005) Disproportionate participation by age/sex classes in aggressive interactions between Long-tailed macaques (*Macaca fascicularis*) and human tourists at Padangtegal Macaque Forest, Bali, Indonesia. Am J Primatol 66:197–204

Fuentes A, Kalchik S, Gettler L, Kwiatt A, Konecki M, Jones-Engel L (2008) Characterizing human-macaque interactions in Singapore. Am J Primatol 70:879–883

Geissmann T, Nijman V, Dallmann R (2006) The fate of diurnal primates in southern Sumatra. Gibbon J 2:18–24

Gillespie TR (2006) Noninvasive assessment of gastrointestinal parasite infections in free-ranging primates. Int J Primatol 27:1129–1143

Goldberg TL, Gillespie TR, Rwego IB, Wheeler E, Estoff EL, Chapman CA (2007) Patterns of gastrointestinal bacterial exchange between chimpanzees and humans involved in research and tourism in western Uganda. Biol Conserv 135:511–517

Gopalan S (1973) Outlines of Jainism. Wiley Eastern Pvt. Ltd., New Delhi

Gössling S (2002) Global environmental consequences of tourism. Glob Environ Chang 12:283–302

Hanson E (2002) Animal attractions: nature on display in American zoos. Princeton University Press, Princeton

Hill DA (1999) Effects of provisioning on the social behavior of Japanese and rhesus macaques: implications for socioecology. Primates 40:187–198

Holmes JC (1996) Parasites as threats to biodiversity in shrinking ecosystems. Biodivers Conserv 5:957–983

Honey M (1999) Ecotourism and sustainable development: who owns paradise? Island Press, Washington, DC

Hsu MJ, Kao CC, Agoramoorthy G (2009) Interactions between visitors and Formosan macaques (*Macaca cyclopis*) at Shou-Shan Nature Park, Taiwan. Am J Primatol 71:214–222

IUCN (2010) IUCN red list of threatened species. Version 2010.4. http://www.iucnredlist.org. Accessed 5 May 2011

Jacobson SK (2010) Effective primate conservation education: gaps and opportunities. Am J Primatol 72:414–419

Jones-Engel L, Engel GA, Schillaci MA, Babo R, Froehlich J (2001) Detection of antibodies to selected human pathogens among wild and pet macaques (Macaca tonkeana) in Sulawesi, Indonesia. Am J Primatol 54:171–178

Jones-Engel L, Engel GA, Schillaci MA, Rompis A, Putra A, Suaryana KG, Fuentes A, Beer B, Hicks S, White R, Wilson B, Allan JS (2005) Primate-to-human retroviral transmission in Asia. Emerg Infect Dis 11:1028–1035

Kindaichi K (1949) Concepts behind the Ainu Bear Festival (Kumamatsuri). Southwest J Anthropol 5:345–350

Kleiman DG, Beck BB, Dietz JM, Dietz LA, Ballou JD, Coribra-Filho AF (1986) Conservation program for the golden lion tamarin: captive research and management, ecological studies, educational strategies and reintroduction. In: Benischke K (ed) Primates: the road to self-sustaining populations. Springer, New York, pp 959–979

Koontz FW (1997) Zoos and in situ primate conservation. In: Wallis J (ed) Primate conservation: the role of zoological parks. American Society of Primatologists, Norman, pp 63–82

Little KA, Sommer V (2002) Change of enclosure in langur monkeys: implications for the evaluation of environmental enrichment. Zoo Biol 21:549–559

Lucas PW, Corlett RT (1998) Seed dispersal by long-tailed macaques. Am J Primatol 45:29–44

Lutgendorf PA (1994) My Hanuman is bigger than yours. Hist Relig 33:211–245

Mallapur A, Choudhury BC (2003) Behavioral abnormalities in captive non-human primates. J Appl Anim Welf Sci 6:275–284

Mallapur A, Sinha A, Waran N (2005) Influence of visitor's presence on the behaviour and welfare of captive lion-tailed macaques (*Macaca silenus*) housed in Indian zoos. Appl Anim Behav Sci 94:341–352

Marriner LM, Drickamer LC (1994) Factors influencing stereotyped behavior of primates in a zoo. Zoo Biol 13:267–275

McCarthy MS, Matheson MD, Lester JD, Sheeran LK, Li J, Wagner SR (2009) Sequences of Tibetan macaque (*Macaca thibetana*) and tourist behaviors at Mt. Huangshan, China. Primate Conserv 24:145–151

Mench JA, Mason GJ (2000) Behaviour. In: Appleby MC, Hughes BO (eds) Animal welfare. CAB International, Oxford, pp 127–142

Mishra C, Madhusudan MD, Datta A (2006) Mammals of the high altitudes of western Arunachal Pradesh, eastern Himalaya: an assessment of threats and conservation needs. Oryx 40:29–35

Mootnick AR, Baker E (1994) Masturbation in captive Hylobates (gibbons). Zoo Biol 13: 345–353

Morgan K (1987) Religion of Hindus. Motilal Banarsidass, Delhi

Moss A, Esson M (2010) Tourist interest in zoo animals and the implications for collection planning and zoo education programs. Zoo Biol 29:715–731

Muehlenbein MP, Martinez LA, Lemke AA, Ambu L, Nathan S, Andau P, Sakong R (2010) Unhealthy travelers present challenges to sustainable primate ecotourism. Travel Med Infect Dis 8:169–175

Nakashima Y, Lagan P, Kitayama K (2008) A Study of Fruit–Frugivore Interactions in Two Species of Durian (Durio, Bombacaceae) in Sabah, Malaysia. Biotropica 40:255–258

O'Leary H, Fa JE (1993) Effects of tourists on Barbary macaques at Gibraltar. Folia Primatol 61:77–91

Orams MB (2002) Feeding wildlife as a tourism attraction: a review of issues and impacts. Tour Manage 23:281–293

Patzschke E, Engelhardt A, Gurmaya KJ (2000) Influences of tourists on the behavior of free-ranging long-tailed macaques (*Macaca fascicularis*). Folia Primatol 71:236

Pedersen BL (1997) Agonistic behavior between rhesus monkeys (Macaca mulatta) and humans in Nepal. MSc dissertation, San Jose State University, San Jose

Phumsathan S, Nepal S (2008) Environmental impacts of tourism, Khao Yai National Park, Thailand. Report. Texas A&M University, Austin

Ram S, Venkatachalam S, Sinha A (2003) Changing social strategies of wild female bonnet macaques during natural foraging and on provisioning. Curr Sci 84:781–790

Randall A (1987) Resource economics, 2nd edn. Wiley, New York

Reade LS, Waran NK (1996) The modern zoo: how do people perceive zoo animals? Appl Anim Behav Sci 47:109–118

Regmi GR, Kandel K (1998) Population status, threats and conservation measures of Assamese macaque (*Macaca assamensis*) in Langtang National Park, Nepal. Report. Primate Society of Great Britain, London

Ross S, Wall G (2001) Wallace's line: implications for conservation and ecotourism in Indonesia. In: Harrison D (ed) Tourism and the less developed world: issue and case studies. CAB International, Wallingford, pp 223–233

Ruesto LA, Sheeran LK, Matheson MD, Li J, Wagner RS (2010) Tourist behavior and decibel levels correlate with threat frequency in Tibetan macaques (*Macaca thibetana*) at Mt. Huangshan, China. Primate Conserv 25:99–104

Savage A, Guillen R, Lamilla I, Soto L (2010) Developing an effective community conservation program for cotton-top tamarins (*Saguinus oedipus*) in Colombia. Am J Primatol 72:379–390

Schnitger FM (1938) The archaeology of Hindoo Sumatra. Int Arch Ethnogr 35:1–44

Sha JCM, Gumert MD, Lee BPY-H, Jones-engel L, Chan S, Fuentes A (2009) Macaque–human interactions and the societal perceptions of macaques in Singapore. Am J Primatol 71: 825–839

Shepherd CR (2010) Illegal primate trade in Indonesia exemplified by surveys carried out over a decade in North Sumatra. Endanger Species Res 11:201–205

Strier KB (2010) Long-term field studies: positive impacts and unintended consequences. Am J Primatol 72:772–778

Supriatna J (2006) Conservation programs for the endangered Javan gibbon (*Hylobates moloch*). Primate Conserv 21:155–162

Taylor P (2003) Goddess on the rise: pilgrimage and popular religion in Vietnam. University of Hawaii Press, Honolulu

Tsujino R, Yumoto T (2009) Topography-specific seed dispersal by Japanese macaques in a lowland forest on Yakushima Island, Japan. J. Anim. Ecol 78:119–125

UNTWO (2011) Tourism highlights. Report. World Tourism Organization, Madrid. http://mkt.unwto.org/en/content/tourism-highlights. Accessed 24 Oct 2011

UNWTO (2010) A year of recovery. Report. World Tourism Organization, Madrid. http://unwto.org/en/publication/annual-report-2010. Accessed 24 Oct 2011

Van Lavieren E, Wich SA (2009) Decline of the Barbary macaque *Macaca sylvanus* in the cedar forest of the Middle Atlas Mountains, Morocco. Oryx 44:133–138

Wallis J, Lee DR (1999) Primate conservation: the prevention of disease transmission. Int J Primatol 20:803–826

Western D (1993) Defining ecotourism. In: Lindberg K, Hawkins DE (eds) Ecotourism: a guide for planners and managers. The Ecotourism Society, North Bennington, pp 7–11

Wheatley BP, Harya Putra DK (1994) Biting the hand that feeds you: monkeys and tourists in Balinese monkey forests. Trop Biodivers 2:317–327

Wiese RJ, Hutchins H (1997) The role of North American zoos in primate conservation. In: Wallis J (ed) Primate conservation: the role of zoological parks. American Society of Primatologists, Norman, pp 29–42

Wolfe ND, Escalante AA, Karesh WB, Kilbourn A, Paul-Murphy J (1998) Wild primate populations in emerging infectious disease research: the missing link? Emerg Infect Dis 4:149–158

Wolfe ND, Switzer WM, Carr JK, Bhullar VB, Shanmugam V, Tamoufe U, Prosser AT, Torimiro JN, Wright A, Mpoudi-Ngole E, McCutchan FE, Birx DL, Folks TM, Burke DS, Heneine W (2004) Naturally acquired simian retrovirus infections in central African Hunters. Lancet 363: 932–937

Wood W (1998) Interactions among environmental enrichment, viewing crowds, and zoo chimpanzees (*Pan troglodytes*). Zoo Biol 17:211–230

WTO (1995) Collection of tourist expenditure statistics. Report. World Tourism Organization. http://pub.unwto.org/WebRoot/Store/Shops/Infoshop/Products/1034/1034–1.pdf. Accessed 24 Oct 2011

Zhao QK (1999) Responses to seasonal changes in nutrient quality and patchiness of food in a multigroup community of Tibetan macaques at Mt. Emei. Int J Primatol 20:511–524

Chapter 7
Pets, Property, and Partners: Macaques as Commodities in the Human-Other Primate Interface

Agustin Fuentes

7.1 Introduction

Until the last decade, most examinations of human-other primate interactions were treated almost exclusively as conflictual and competitive. Crop raiding and related resource exploitation patterns by primates were the focal point for the study of human-primate interactions (e.g., Paterson and Wallis 2005). The human-other primate interface was usually examined in the context of competition for space and resources and as a direct contest or conflict between humans and other primates, with the other primates usually winding up as the losers. However, while such competition does occur, it is an incomplete description of the interface, and the coecologies, of humans and other primates (e.g., Fuentes and Hockings 2010; Riley and Priston 2010).

Two recent arenas of research challenge the traditional approach to this interface: ethnoprimatology and multispecies anthropology. The ethnoprimatological approach focuses on the role of humans as primates, of other primates as coparticipants in shaping social and ecological space, and integrates methodologies from primatology, anthropology, and other approaches in understanding the human-other primate interface (Fuentes 2012; Hockings 2010; Riley 2007). The multispecies approach is seen in the emergence of "multispecies ethnographers" who emphasize the subjectivity and agency of organisms whose lives are entangled with humans. Kirksey and Helmreich (2010, p. 566) tell us that "Multispecies ethnography emerged at the intersection of three interdisciplinary strands of inquiry: environmental studies, science and technology studies (STS), and animal studies. Departing

A. Fuentes (✉)
Department of Anthropology, University of Notre Dame,
Notre Dame, IN 46556, USA
e-mail: afuentes@nd.edu

S. Radhakrishna et al. (eds.), *The Macaque Connection: Cooperation and Conflict between Humans and Macaques*, Developments in Primatology: Progress and Prospects 43,
DOI 10.1007/978-1-4614-3967-7_7, © Springer Science+Business Media, LLC 2013

from classically ethnobiological subjects, useful plants and charismatic animals, multispecies ethnographers also brought understudied organisms—such as insects, fungi, and microbes—into anthropological conversations."

These two intertwined approaches force us to shift our gaze from the conflict at species interfaces to encompass the mutual ecologies and histories, viewing these interfaces as interweavings of the social and structural ecologies of the participants into a nature-cultural relationship (Fuentes 2010). This approach is thus concerned with the integration, engagement, and interface between ourselves and other kinds of life (Kirksey and Helmreich 2010; Kohn 2007). Viewing relationships between organisms as shaping and being shaped by mutual ecologies in behavioral, ecological, and physiological senses, we can make greater inroads into our interdisciplinary conversations about the Anthropocene (Rose 2009). Using ethnoprimatological and multispecies perspectives as the frame for studying the human-other primate interface can enhance our abilities to envision emergent relationships in the Anthropocene, wherein human social, political, perceptual, and economic actions are entangled with other primates' (and other organisms') behavioral and ecological lives (Fuentes 2010; Kirksey and Helmreich 2010).

In this chapter, I will focus this broader lens on specific aspects of the interface between macaques and humans and make the argument that in some of these cases macaques can be described as specific types of commodities for the human interactors, beings whose presence and/or use can provide "added good" in a social and/or financial sense (e.g., Fuentes 2010; Knight 2006; Ohnuki-Tierney 1987). I suggest that in many ways our understanding of the macaque-human interface can benefit if we conceptualize macaques, in certain contexts, as social commodities, financial commodities, and cultural commodities.

7.2 Humans Often Negatively Impact Other Primates, But It Is Not Always so Simple with the Macaque-Human Interface

There is general agreement that human impact via population expansion, forest modification, and landscape development generally has a negative impact on primates. Through diminishing and restructuring range areas, there is a continuously increasing likelihood of human-primate contact and conflict, with the nonhuman primates usually on the losing end of such interactions (Hockings and Humle 2009; Campbell-Smith et al. 2010; Hockings et al. 2010). However, anthropogenic modifications can also enhance pathways between groups in a population of primates, especially macaques. For example, the majority of macaques on Bali, Indonesia, and across Thailand are found at sites associated with temples and receive significant components of their nutritional requirements from humans or human activity (Lane et al. 2010; Gumert 2011). In many cases, land-use patterns combined with the complex temple and religious landscapes create forest refugia and corridors that fit remarkably well with the macaques' patterns of habitats use, foraging, and dispersal (Fuentes et al. 2005; Lane et al. 2010; Gumert 2011). In these cases, macaques are actually benefiting from such anthropogenic patterns, which sets the stage for seeing the macaque-human interface as somewhat distinct from that of most other primates.

Primates are hunted by many human cultures. Hunted primates can be food sources or to satisfy a variety of human cultural needs ranging from ingredients in traditional medicines, through social pets, to subjects of biomedical research. Long-term, sustained predation by humans at a level below extinction-causing may impact the behavior and morphology of nonhuman primates, and it is also possible that selective hunting for body parts or other features that are used in traditional medicines can negatively impact primate populations. However, in the case of macaques, hunting them as food is neither ubiquitous across cultures, nor are macaques usually the favored food species in areas where they are sympatric with other primates (e.g., Fuentes 2002). In fact, much hunting of macaques is not for direct consumption but rather for the capturing of the macaques for resale either as pets or for biomedical research use. As with anthropogenic land use, the human-macaque interface surrounding hunting also differs to a degree in pattern and outcome from that of other human-primate relationships.

Our understandings of the patterns and contexts of shared pathogen environments and their evolutionary implications remain incomplete; however, we know that humans and macaques are particularly mutually interactive in this arena (Engel and Jones-Engel 2011). Close contact and range overlap between humans and primates do introduce the context for disease transmission (Jones-Engel et al. 2008; Wallis and Lee 1999), and throughout their range, most macaque species share a much higher degree of overlap with humans than other primate species. In general, researchers have shown that humans interacting with macaques risk exposure to a number of simian viruses, including simian T cell lymphotropic viruses (STLV), simian retrovirus (SRV), simian foamy virus (SFV), and herpes B virus in addition to other known infectious agents (Jones-Engel et al. 2008). Macaques can also contract human pathogens, including measles, influenzas, and other respiratory pathogens such as parainfluenza and tuberculosis (Jones-Engel et al. 2001). Interestingly, both humans and macaques can act as reservoirs for shared pathogens or potential pathogens that move across species boundaries, such as with malarias (Cox-Singh and Singh 2008).

I, and many others, have argued that primate studies must move toward an ethnoprimatological and multispecies approach with explicit acknowledgment of the multifarious nature of the human-other primate interface. This repositioning moves us beyond the basic paradigm of conflict and competition (Fuentes 2012; Hockings 2010). Rather than focusing primarily on the behavior and ecology of the primate species at hand, or on symbolic interpretations/uses of primates, multispecies ethnoprimatology attempts to merge these two approaches, and others, into a broader and more integrative framework. Using such an approach, we can more effectively move beyond standard representations and consider the array of ways in which macaques can be considered simultaneously commodities in a human system and participants in a broader multispecies ecology.

7.3 Macaques as Commodities?

What is the benefit to conceptualizing and studying the macaque-human interface through multiple lenses? First of all, it recognizes reality. Traditional primatological approaches focus on behavioral ecology and ignore, or only partially incorporate, significant anthropogenic elements in local and global ecosystems. While they do provide us with absolutely important data on ranging and feeding, social behavior, and kin relationships, they tend to treat the lives and niches of the macaques as primarily separate from the human societies around them. In the modern context of human-macaque overlap, this paints a woefully incomplete picture.

Most primatologists would agree that adding genetics/genomics, plant and soil ecology, and multispecies use/partitioning of the local ecologies to studies of macaque groups and populations would benefit our understanding. Those interested in ethnoprimatology would also add human land-use patterns and the patterns and outcomes of the human-macaque interactions, and possibly human perceptions of the macaques and the history of macaques in the local human society. Ethnoprimatologists argue that these factors are as important facets of the local ecosystems as are habitat structure and local phenology. Here I suggest, along with many others (e.g., Fuentes 2002; Knight 2006; Riley et al. 2011), that we can add additional levels of conceptual analyses to studying macaque-human interfaces. Specifically, I propose we consider that macaques, in some contexts, can act as specific types of commodities in human societies. If this is the case, then a fuller understanding of the macaques' behavior, ecology, and histories requires an investigation of the ways in which their "commodification" by humans might affect the macaque's landscapes, ecologies, and the niches in which they exist.

7.3.1 Why the Concept of "Commodities"?

Many animals are already envisioned, in an economic and historical sense, as commodities for human beings. Here, a commodity is defined as an article of trade or commerce and as a product as opposed to a service. More generally, commodities (animals and things) are seen as something of use, advantage, or value to a given society, something whose presence and/or use can provide added "good" in a social and/or financial sense. For our consideration then, a commodity is an object or item that is worth some measure of value to the humans having it, encountering it, and coexisting with it. I would like to suggest that in many (but not all) of the cases where humans and macaques coexist, especially in South and Southeast Asia, macaques may be acting as commodities for the humans. In these cases, the macaques are, more than other animals or beings, sharing the human environment, and they are objects with worth in the social and structural economy.

However, I am not implying that we should use a market approach to assess human-macaque interfaces. Many conservation projects have initiated valuation of particular species over others (invasive versus endemic, pest damage to crops, predators versus prey, etc.); this is not my intent. I am not arguing for a shift to a market-based ethnoprimatology. Rather, I use the concept of commodity to reflect a way to assess and envision the underling perceptive context for investment by humans and its concomitant impact on the macaques' behavior, landscapes, and ecologies. I do not suggest that an economic cost-benefit approach to valuating the interspecies interface is beneficial.

This is not to say that there have not been previous works looking at the economic and/or social values of macaques as central to coresident human perspectives (e.g., Knight 2006; Sponsel et al. 2002). Here, I hope to add to this literature and argue that more primatological projects could benefit by adding the perspective that the macaques they study have some economic or social value to the humans around them and thus the landscape, the macaques ecology, will be structured by this aspect of human perception. I also want to point out that here I am suggesting more than simply viewing macaques as having dietary or monetary value (i.e., meat for sale). The idea that the macaques mean something more than meat, or a pest, or just another local animal to the coresident human populations expands our focus such that a multispecies and ethnoprimatological context becomes required for effectively assessing the lives of the macaques (for earlier version of this call, see Fuentes 2002; Riley 2007; Riley et al. 2011). This becomes important to those studying macaques when we consider that human perception plays a substantive role in structuring the processes and outcomes of human-other animal interactions. If the macaques are seen, at least in part as financial, social, or cultural commodities, then their relationships with humans will be shaped by that context. This different perception and classification of macaques might set them in a distinct category, or give them a more complex social valuation, relative to other animals in peoples' lives and spaces. If this is the case, then these perceptions can have impacts for conservation and management of macaque populations overlapping with humans. In the next section, I will provide a few brief examples of how macaques can be considered as cultural, social, and financial commodities.

7.4 Macaques as Cultural Commodities

Seeing macaques as a cultural commodity reflects scenarios where macaques have an integral inclusion as an important aspect of a human society; an aspect that is not just food, or something of cash value, but that provides some form of cultural "good." Examples of this type of relationship include monkey performances in Japan (which also occur throughout South and Southeast Asia) and the prominent social inclusion of primates, especially macaques, in mythos and cultural narratives in popular religion and secular contexts.

Monkey performance is a blend of the economic and social in that there is a financial underpinning for the monkey trainer but the content and context of the performance relies on a specific set of assumptions/beliefs about the macaques who are the performers. Monkey performances include a variety of theatrical performances by trained macaques, where the monkeys mimic specific types of human cultural behavior and social scenarios via staged interactions with their trainer and the audience. The macaques are often clothed in specific garb that denotes the human social roles they are playing. Generally, the audience provides monetary contributions to the trainer at the conclusion of the performance. These funds are the income of the trainer and are used to maintain both the monkey(s) and the trainer's family. While these performances were until recently widespread across Asia, they are now most common in India and parts of Southeast Asia and parts of Japan. Interestingly, historically these types of performances were common in areas of northern Africa and Europe (Janson 1952), but now are found only in a few areas in Morocco (all with the macaques species *Macaca sylvanus*).

A well-known in-depth study of monkey performance is Emiko Ohnuki-Tierney's work on Japanese macaque (*Macaca fuscata*) performances (Ohnuki-Tierney 1987, 1995). She presents the notion that in Japan the monkey performance, and the macaques themselves, act as a mirror for humanity "playing a powerful role in their (humans') deliberations of who they are as humans vis-à-vis animals and as a people vis-à-vis other peoples" (Ohnuki-Tierney 1995, p. 297). Monkeys are the only animal addressed as "san," the adult human address, and are referred to as "humans minus three pieces of hair." This is a real type of social inclusion: there is a special role for the macaques in Japanese culture that is distinct from the characterizations of other animals (Ohnuki-Tierney 1987, 1995). This role can be seen as serving a value, acting as a social commodity, for the Japanese, while simultaneously acting as a financial commodity for the monkey trainers.

The yaen ko¯en or "wild monkey parks" of Japan might be seen as having grown out of the tradition of monkey performance. These parks are significant tourist revenue generators and have substantial impacts on the ranging and overall ecology of macaques in many parts of Japan (Knight 2006). In these areas, management teams provision free-ranging macaques such that they influence the ranging patterns and create a zone of habituation where macaques and humans overlap: macaques eat and interact with one another, while humans watch them and value their exposure to significant animal others in what they perceive as a natural setting. These "natural zoos" are the result of a Japanese social valuation of seeing "wild" monkeys up close and the ability of management teams to generate substantial economic returns from controlling the monkeys' ranging behavior for the benefit of the humans coming to see them (Knight 2006). The integration of economic and cultural roles for macaques in Japan reveals a significant cultural relationship between humans and monkeys even when conflict over land and crops is seen as a core problem between humans and free-ranging macaques in Japan (Sprague 2002). While the Japanese case is well studied, there are similar scenarios involving monkey performance throughout South Asia as well.

In addition to monkey performance, the role of macaques in human mythos (both secular and religious) is widespread. For example, humans and macaques have coexisted in Gibraltar, at the southern tip of the Iberian Peninsula, for centuries, with evidence showing a presence of *Macaca sylvanus* since the 1300s (Cortes and Shaw 2006). Today, the macaques play both cultural and economic roles for the Gibraltarians.

A popular British and Gibraltarian tale of the macaque-human alliance on Gibraltar speaks of Spain's last attempt to gain control of Gibraltar by force. According to local lore, Spanish troops ascended the eastern side of the Rock of Gibraltar in order to ambush the British-held port town of Gibraltar on the west face. These troops happened across a group of sleeping macaques, who sent up a series of alarm calls that awoke the British garrison. With this warning, the British troops were able to repel the attacking Spanish, and Gibraltar has been in British hands ever since. The famous British prime minister, Winston Churchill is reported to have declared that as long as the Barbary apes (the local name for *M. sylvanus*) are on the Rock, Gibraltar will remain British and even assigned the British army to care for the macaques.

While today the macaques of Gibraltar serve as a significant financial commodity for Gibraltar (nearly 15% of tourists visiting are there to see the monkeys), they also act as a significant cultural commodity for the Gibraltarians. The macaques bring a certain "environmental" status (and its political implications) to Gibraltar by making it the only location with a free-ranging, native, nonhuman primate species in the entire European Union. The macaques also act as a symbol for the British claim to the city and space of Gibraltar. This claim is much disputed by the Spanish, who held dominion over the area for nearly 300 years prior to the British (who have now held Gibraltar for over 300 years). The macaques are a type of commodity, a symbol that has a social value in the broader political debate between countries about boundaries, environment, and dominion.

With regard to macaques across Asia, two specific socioreligious traditions are especially relevant to inquiries about social and cultural commodities: the *Ramayana* and the *Journey to the West*. The *Ramayana* is a major Hindu epic more than 2,000 years old. A major character in the *Ramayana* is Hanuman, the monkey king/general who is a loyal and devout servant of Rama, the central hero of the epic. Hanuman and his monkey army help in the rescue of Rama's bride, Sita, from the demon lord of Lanka (Ravana) and the defeat of the demon armies. Because of the substantial impact of Hindu cultures across the region, this epic and a wide array of affiliated folktales are prominent throughout South and Southeast Asian societies, even in those that are not primarily Hindu today. In many of these societies, Hanuman and his monkey legions (often depicted as macaques) are considered to have a positive mythical value. This results in many monkeys species (frequently macaques) being tolerated or even provisioned in and around temples and other Hindu and Buddhist shrines (Riley et al. 2011). Given the sincere devotion of Hanuman to Rama and the role that he played in assisting "man" combined with long-term exposure to monkeys by the human populations in South/Southeast Asia, we can see that macaques can be seen as a symbolic good in folkloric contexts, which translates to substantive outcomes for the humans and the macaques across much of the region (see also other chapters in this book).

The Chinese novel *Journey to the West* dates to the sixteenth and seventeenth centuries, but appears to have older roots. This novel is a confluence of Daoistic, Confucianistic, and Buddhistic perspectives and might also be influenced by the *Ramayana*. The novel opens with the birth of monkey (Sun Wukong), who becomes the king of the monkeys. He winds up creating chaos in heaven as he challenges the pantheon of deities, upsetting the Jade Emperor. He is imprisoned beneath the Mountain of the Five Elements by Buddha. Subsequently, he is released to accompany a monk on his quest to obtain the holy Buddhist scriptures from India. On the journey monkey helps the monk overcome an array of challenges, supernatural obstacles, and monsters before reaching India and returning to China with the Buddhist texts. The story is loosely based on an actual journey undertaken by the monk Xuanzang (600–664 AD) to India, the home of Buddhism, to collect Mahayana Buddhist texts for translation into Chinese (Burton 2002). While the monk is the ostensible focal point of the story, it is the charm, guile, and exuberance of the monkey that often reads as the central entertainment of the novel. In the end, Sun Wukong (monkey) achieves a form of enlightenment, earning the title "Buddha victorious in strife" (Burton 2002). This tale is extremely popular throughout much of Asia, especially China. The popularity of this story impacts the conceptualization of primates in many of the human cultures where it is told. The role of monkey as assistant to humanity, often engaging in chaotic and hierarchy challenging behavior at the same time, allows a symbolic and mytho-historical context for the interaction patterns between humans and nonhuman primates (especially macaques) in much of Asia. This contextualizes (folklorically) the food-raiding behavior, the monkey performances, and other aspects of the human-macaque interface. The story of a transition to maturity allows for the cultural role of monkey as teaching tool, and therefore as a valuable cultural commodity, to emerge. This emergence enables a more extensive overlap, and incorporation, of macaques into, and with, human ecologies than other primates.

These examples suggest that there are specific ways in which macaques are interwoven into human mythos and folklore. I suggest that these patterns create human perception of the macaques as contributing to the social good of humans and that humans can see a benefit to being around macaques and having them as part of a larger multispecies association. This status is an important aspect of human social and structural ecologies that we must recognize and examine when thinking, and constructing explanations, about macaque ecologies, behavior, and ranging. This by no means suggests that there is not also conflict and competition between macaques and humans in these same areas. It is just that the overall relationship is greater, and more complex, than the general conflictual interfaces between humans and other primates and between humans and other large nondomestic mammals.

7.5 Macaques as Partners and Financial Commodities: Monkey Tourism and Coconut Pickers

Macaques are used as economic tools in many areas across Asia. However, these uses are not only classifiable as foodstuffs or other standard units of exchange. Rather, macaques, in some cases, are better seen as economic collaborators (as in temple sites) or as "laborers" for humans in a particular type of production (akin to certain domestic animals such as oxen). These types of economic relationships, where macaques serve as active and collaboratory financial commodity, are quite distinct from most other primates' relationships to humans.

Coconut picking is an excellent example of macaques as economic tool and "labor" for humans. While there are references to coconut-picking macaques across much of southern and Southeast Asia, the relationship has been studied most extensively in Thailand where male pigtailed macaques (*M. nemestrina*) are kept, raised, and trained to pick coconuts (Sponsel et al. 2002). Humans in southern Thailand capture macaques as young individuals and then train them to be effective coconut pickers. The investment of time and energy that goes into training the macaques is intensive, but the payoff can be significant. A well-trained macaque can harvest between 500 and 1,000 coconuts per day and with a lower cost and higher return than a human could (most of the coconut picking is still done by humans). Examining this system from primatological and ethnographic perspectives, Sponsel et al. (2002) suggest that this is a cooperative human-monkey economic relationship. They argue that in certain areas of Thailand cultural and agricultural systems facilitated a move from conflict between humans and macaques over crops as the dominant relationship toward a synergistic relationship wherein the humans capture, train, and maintain the macaques as they serve their economic role (as coconut picker). Obviously, there remains conflict between humans and macaques in these areas, and this system results in some small number of male macaques being removed from local populations. However, it is a particular type of economic relationship that will affect human perceptions of the macaques and impact local economies and ecologies.

At many temple sites across south and southeast Asia, macaques and human coexist in high densities for long periods of time. Previously, explanations for this type of relationships have focused on connections between the resident macaque species and the local humans, invoking the monkeys' role in local mythos or religious folklore as setting the stage for a beneficent view of the macaques. However, it is quite possible that in these cases it is often the macaques' role as economic commodity combined with their cultural commodity status that enables a toleration and even kindness by humans toward resident macaques (Fuentes et al. 2005; Fuentes 2010). This creates a scenario in which macaques in and around temples are more likely to receive food, protection, and consideration from humans. Their role as cultural commodities sets the stage for the occurrence of partnership between macaques and humans that creates the context for macaques to emerge as financial commodities in these temple contexts.

Many Hindu and Buddhist temple sites have macaques associated with them, and some of these sites have become significant tourist attractions, destinations often referred to as "monkey temples" in travel books, official tourist campaigns, and via tour agencies and guides. Such temples can be found across India and South Asia, in mainland Southeast Asia (especially Thailand and Malaysia), and in Indonesia (especially Bali). In Southeast Asia at these temple sites, tourists either pay to enter the temple area or they purchase food for the macaques from local vendors (or both). The income from entrance fees can quite significant for local populations and is often used in the upkeep and maintenance of the temple grounds. Also, in some instances, local humans and the macaques collaborate to extract cash from tourists. Macaques take items (sunglasses, earrings, bags, etc.) from the tourists, and local humans approach and retrieve the items (by giving food to the macaques) and receive a small amount of money from the grateful tourists in return. While not actually a "prearranged" collaboration, the result is one where both the humans and the macaques have a common input into a system that benefits each economically (money for humans and food for the macaques).

This relationship is exemplified by the pattern of interactions and coexistence between *Macaca fascicularis* and humans on the island of Bali, Indonesia (Fuentes 2010). The macaques on Bali are found throughout the island, except in the highly urbanized regions. There are at least 63 sites where macaques reside on the island (Fuentes et al. 2005). Each of these sites has one or more groups of macaques that use the site, and more than two-thirds of these sites are associated with a temple or shrine. These religious complexes can be as small as a simple shrine consisting of a few stones and an altar to elaborate temple complexes that are heavily used by Balinese and, in some cases, foreigners. Macaque groups undergo some provisioning as sites that contain a shrine or temple complex have temple ceremonies several times per year and many of the sites also provide food to the macaques via direct provisioning and/or tourist provisions. Tied to this food availability, temple site populations tend to have larger group sizes than non-temple site populations (Lane et al. 2010).

There are at least 10–15 of these temple sites in Bali that are heavily visited by tourists, sometimes as many at 100–120,000 per year. Entrance fees range, but are generally in the USD 1–3 or equivalent range. All of these temples are ringed by local vendors' stalls, many have local tour guides, and guided tours from across the island bring tourists to the sites. At these temples, the Balinese gain specific economic benefits from the presence of the macaques, and the macaques gain both nutrition and protection from the relationship: this scenario reflects macaques as both cultural and financial commodities (Fuentes 2010; Wheatley 1999).

7.6 Macaques as Pets: Social and Financial Commodities

"Pets" is a substantial category. It can be animals kept for companionship, enjoyment, or status in addition to those that contribute to the income, nutritional intake, or other functional facets of the humans who own them. Today pet ownership of a few macaque species (primarily *M. fascicularis* and *M. mulatta*) is widespread in Southeast Asia and occurs in South Asia, while in Southeast Asia, *Nycticebus* (a prosimian, the slow loris) is also a common pet (Malone et al. 2003; Nijman 2009; Shepherd 2010). It is not clear how the capturing and keeping of pets impacts wild populations of macaques. In all likelihood, pet keeping probably functions in a similar fashion to predation, as pets are frequently by-products of hunting, and capture for pet keeping probably affects populations in a way similar to, or greater than, extraction via predation (Eudey 1987, see also Nijman 2009). Because the pet tradition is quite old in Southeast Asia, it is also quite likely that there is substantial bidirectional pathogen transmission between humans and their primate pets (Jones-Engel et al. 2001). It is possible that there are physiological changes, even adaptations, in populations of human and non-human primates that overlap extensively, such as pets, humans, and the local populations that the pet primates come from, as a result of these intensive interactions. Thus, the "pet" relationship can be an important component in the ecologies of humans and macaques. We can divide the category of pet into three major types: social pets, status pets, and food pets. In the case of macaques, the first two are the most common, and all three categories can be seen as types of commodities.

Social pets are those macaques that are kept much in the same way that many peoples keep dogs in the western world. A household might have one or more macaques as playmates for children and/or for general companionship in the household. In my travels throughout Southeast Asia, I have noted that these pets are generally kept tethered in common areas of the house or in a courtyard or other open area associated with the house. The tethering is usually by a belt around the waist and long cord or rope. As the macaques age, if the belt is not adjusted, abrasions leading to infections and abscesses are common. The pattern of older pet macaques suffering injuries from their tethers is quite common and might be a significant factor in pet mortality (at least in social and status pets).

When the macaques are infants or young juveniles, they may be carried on the shoulders of members of the household or carried in the arms, even outside of the house. Anecdotal evidence from Indonesia and Thailand suggests that this type of contact is more common in rural areas than in urban ones, indicating that rural pet owners may have more (or more regular) physical contact with their pet macaques than urban pet owners (L. Jones-Engel and N. Malone, 2002). Most pet macaques are juvenile individuals (less than 3 years of age), and many are released (or slaughtered, see below) as they reach adulthood. As the monkeys become adults, they are more likely to act aggressively toward humans, and they are also significantly stronger. Part of this aggression may be related to ill health associated with types of housing/caging and/or infection from their tethers/restraints (Malone et al. 2003; Shepherd 2010). Alternatively (or additionally), this aggression might reflect normative behavioral shifts elicited by the coincidence of captivity and the onset of adult hormone physiology. In the case of males, the canines increase in size as the animal matures, adding to the risk of aggressive behavior.

If pets are released, it is often into forested areas where free-ranging macaques are known to reside or at temple complexes with existing macaque populations. There are no studies (that I know of) that have either documented release patterns or followed released macaques into forests or temple-based groups; thus, we have no real sense of the relative mortality in these cases. One interesting example suggesting some degree of success in such releases is the case of Hong Kong. Over the 1900s, there were a number of released macaques from at least two species (*M. mulatta* and *M. fascicularis*), many of whom were able to form groups, accept some continued inflow of released animals, and even breed across the species. The Hong Kong macaques now number over 2,500 in multiple groups. It is possible that the physiological resilience of macaques combined with their general pattern of male migration establishes a baseline wherein former pets, if they are males, have some chances of integrating into free-ranging or semi-free ranging groups (see Wong 2000).

In many of the cultures where macaques are kept as pets, there has not been strong social dog or cat ownership tradition (although that is changing in many urban areas). In these societies, dogs are frequently kept as household guards or watchers, but the high frequency of social and physical contact is not common. It is likely that in these societies macaque pets have served a somewhat similar role as dogs do in the west (part-time social companion and potential stress relief). Regardless of the duration of their stay within the human household, these pets may be acting as a temporary social good within that household, potentially providing stress relief, entertainment, and social companionship.

Status pets are kept as indicators of the affluence of their owners. It is not overly common for primates to be status pets, except for apes (such as orangutans). However, in some areas of Southeast Asia (e.g., Indonesia, Thailand, and southern China), ownership of multiple "wild" or exotic animals can be an indicator of status and wealth. While most of the animals are birds, reptiles, and fish, some mammals, including macaques, are also occasionally found in these menageries. This is most frequent with macaques that are not native to the region in which they are kept or that are less common and/or that have uncommon pelage colors or patterns (Malone et al. 2003, see also Nijman 2009 for gibbons and orangutans).

Status pets are less likely to be the ubiquitous *M. fascicularis* and more likely to be pigtailed macaques (*M. nemestrina*) or other species less frequently found in associations with human settlements. Unlike the social pets, these individuals are often caged or tethered in a more restrictive and less interactive manner. They generally have limited social interactions and often exhibit stereotyped, repetitive behaviors and engage in self-injurious behavior. These pets are generally purchased at animal markets as juveniles or infants, and in many cases, they have a variety of maladies (skin parasites and respiratory illnesses are common) and are often in generally poor health (Jones-Engel et al. 2001; Malone et al. 2003; Shepherd 2010). Status pets are general object commodities serving as indicators of status for the owners; their value lies in the fact that animal ownership demonstrates a degree of disposable income and thus reflects a social and economic status. In certain cases I have observed, particularly in parts of Indonesia, there is a conceptualization of "good luck" associated with owning animals, so the status provided is not necessarily reflective of wealth but rather an investment by the owner in strategy to ensure auspicious outcomes.

Finally, food pets are a temporary class of pet that is kept only until it reaches a particular age or size when it is then eaten. This is primarily found in rural populations and often results from a hunt where the mother was captured and consumed and the hunters kept the infant. In this case, there may be some overlap with social pets as children and others may care for and play with the macaques as it grows. However, this is always short-lived as the pet is eventually consumed. With macaques, this is probably the least common form of pet status; however, it is clearly a direct form of nutritional commodity similar to a domestic chicken or pig raised for food (see Fuentes 2002; Jones-Engel et al. 2001).

In each of these categories, there is some possibility of pathogen transfer between the pet and the owners. Documentation of the transfer of nonpathogenic retroviruses

(SFV) and of other simian pathogens suggests that pets and owners have a higher likelihood of such exchanges. However, it is not clear what sort of long-term impacts this might have on the human populations (at least in the case of macaques and humans). There is no evidence that macaque pet owners are contracting herpes B, even though a larger percentage of pet macaques who have been sampled appear to have the virus (as do most adult macaques). The recent multiple discoveries that *Plasmodium knowlesi* has a history of transfer from macaques to humans reinforce the notion that these kinds of pathogen exchanges are relatively common and might be accentuated via pet ownership (Engel and Jones-Engel 2011).

Finally, it is worthwhile to note that in most of the countries where humans do keep macaques as pets, the practice (keeping a primate as a pet) is illegal. In most cases, this is minimally enforced, but when it is enforced, the police usually confiscate the macaques, sometimes fining the owner, and deliver the macaque to a rescue center or wild animal shelter (Nijman 2009; Shepherd 2010). The disjuncture between social customs of pet keeping and national laws, which are heavily influenced by international primate and animal welfare action and legislation, is relatively recent. It may be that over the last few decades overall macaque pet ownership has been decreasing and may be something that will disappear in the not too distant future. However, there are few surveys and no baseline data to base an assessment of overall pet macaque numbers either in a specific country or across Asia. Describing the patterns of pet keeping, the relative ways in which the macaques are perceived as commodities, and considering the political and economic contexts of the societies where pets are kept can provide a baseline and context for future examinations of macaques as pets.

A final category where macaques can be seen as more traditionally conceptualized financial commodities is worth noting: biomedical capture and use. While India did away with this practice over three decades ago, China, Indonesia, and other countries in Southeast Asia (and Mauritius) continue to capture, breed, and export macaques for invasive and lethal research purposes. In these cases, macaques are explicitly used as simple economic commodities, objects for use and disposal. Many countries in the European Union are reducing or banning this type of research, but the USA and China continue significant research use of macaques (primarily *Macaca fascicularis*). I am not engaging with this context in this chapter, but it is worth noting how this type of economic exploitation of macaques might affect the human perceptions of macaques in both the exporting and importing countries (see Ahuja, this volume).

7.7 Implications of Using This Framework to Examine the Human-Macaque Interface

This chapter is not an argument against traditional approaches in ecology and behavior; these approaches are the cornerstone of primatology. I am merely suggesting that we consider the possibility that understanding the perceptions of macaques' economic, social, and cultural roles in local human societies adds value and depth to studies of macaques. Viewing macaques as commodities, in certain contexts, enables us to see that they play specific types of roles in the lives of humans and that these roles can be important in shaping the behavior and potential future of macaque populations throughout Asia.

One drawback to utilizing the terminology of "commodity" is that it is commonly used in market-based analyses; this is not how I am using it here. I am not comparing macaques to salmon being farmed in New Zealand or to pigs being raised in northern Spain. It is not my intent to argue that we should assess, via a cost-benefit analysis, the financial and social value of macaques in order to evaluate whether the interface is "good" or "bad" in terms of net benefits, or whether or not the macaques should be kept or removed.

Recently, there has been much work envisioning other beings and natural systems as "goods" that perform "ecological services" in an economy of nature. This has been called a "pragmatic ecology" and is emerging as a major approach in ecology and conservation management in the twenty-first century (Palmer et al. 2004). This approach focuses on ecological structure and design within ecosystems relative to human use and alterations. It looks at changes in ecological commerce, the movement of living and nonliving materials that influence ecosystems and ecological processes, and accepts that, in the Anthropocene, humans are key players in nearly every ecosystem on the planet (Palmer et al. 2004). This is an important approach for broad scale ecology and ecosystems management, but one of its primary goals is to maintain and manage systems that have long-term core productive roles for human beings. This orientation is not what I am proposing here. Such a pragmatic ecology can add to our overall understanding of ecosystems and the roles that their components play in the context of outcomes relative to human lives; but this is not the goal of an ethnoprimatological approach to understanding the macaque-human interface. Nor is it the goal of a multispecies approach that extends the point of focus to beings aside from humans, and alongside with humans, in envisioning ecosystems.

I am arguing here that using the concept of commodity as part of an extended toolkit in ethnoprimatology can enable us to leverage the fact that in some contexts, especially in South and Southeast Asia, macaques have specific types of value; they mean something for the coresident humans. Thinking about this "meaning" in terms of social, financial, or cultural contexts helps us dissect the components of this multispecies relationship and apply an ethnoprimatological approach to understanding the overall ecosystem in which macaques and humans coreside and coconstruct. However, this is not the only aspect of the relationship that is important in viewing the lives and ecologies of the macaques. Looking at the human perceptions of macaques in a given locale must also be accompanied by observing the basics of the macaques' behavioral ecology, by basic ethnographic focus on the humans, and by a broad assessment of the local ecologies (e.g., Fuentes et al. 2005; Riley and Fuentes 2011).

Differential perception and classification of macaques relative to other primates and/or mammals can set them in a distinct category relative to other animals, and thus, their presence in peoples' lives and spaces will take on a different trajectory. These trajectories and perceptions can have impacts for conservation and management of macaque populations overlapping with humans. Considering these human perceptions of a social, cultural, and/or economic value of coresident macaques adds a specific item to our toolkit, an aspect of the anthropogenic ecology, in our overall modeling of the macaques' ecology.

Macaques appear to be more successful at coexisting with humans in much of Asia than most other primates. Traditional ecological arguments for macaques as "weed-species" and edge-specialists are not sufficient to explain this. Ethnoprimatological research suggests that it is also the role of the macaque in human societies, and the ways in which humans commodify the macaques, that plays a significant role in the macaques' success around humans (Fuentes 2010; Riley and Priston 2010). Our macaque-focused models can include more than general behavior, food availability, and ranging area; they can include local human cultural customs, social contexts, and market practices as they affect aspects of the macaque populations. Combining traditional approaches with ethnoprimatological ones, in the context of multispecies communities, produces a wider array of hypotheses and potential data sources for our investigations into macaque behavior and ecology.

Finally, if we are to consider the future of macaques at the macaque-human interface, we are faced with a problem: there is conflict between human populations and other large mammals, and human populations will continue to grow. Even macaques suffer increased negative repercussions from this pattern. In areas where there are large populations of macaques coresiding with humans, there is a conflict for space and food at some level, and this will only continue to increase as the growth and spread of humanity continues. This suggests that a multispecies management approach is needed if we are interested in the sustainability of macaque populations in and around human ones. Understanding the value of macaques in terms of their perceived commodification potentially provides insight into ways we can influence the human population in the construction of programs for the maintenance of macaque populations as participants in the local multispecies communities.

References

Burton FD (2002) Monkey king in China: basis for conservation policy? In: Fuentes A, Wolfe LD (eds) Primates face to face: the conservation implications of human-nonhuman primate interconnections. Cambridge University Press, Cambridge, pp 137–162

Campbell-Smith G, Simanjorang HVP, Leader-Williams N, Linkie M (2010) Local attitudes and perceptions toward crop-raiding by orangutans (*Pongo abelii*) and other nonhuman primates in northern Sumatra, Indonesia. Am J Primatol 72:866–876

Cortes J, Shaw E (2006) The Gibraltar macaques: management and future. In: Hodges K, Cortes J (eds) The Barbary macaque: biology, management and conservation. University of Nottingham Press, Nottingham, pp 199–210

Cox-Singh J, Singh B (2008) Knowlesi malaria: newly emergent and of public health importance? Trends Parasitol 24:406–410

Engel G, Jones-Engel L (2011) The role of *Macaca fascicularis* in infectious agent transmission. In Gumert MD, Fuentes A, Jones-Engel L. 2011. Monkeys on the Edge: Ecology and Management of Long-Tailed Macaques and their Interface with Humans. New York: Cambridge Univ. Press 2011, pp. 183–204
Eudey AA (1987) Action plan for Asian primate conservation: 1987–1991. IUCN/SSC Primate Specialist Group, Gland
Fuentes A (2002) Monkeys, humans, and politics in the Mentawai islands: no simple solutions in a complex world. In: Fuentes A, Wolfe LD (eds) Primates face to face: the conservation implications of human-nonhuman primate interconnections. Cambridge University Press, Cambridge, pp 187–207
Fuentes A (2010) Nature cultural encounters in Bali: monkeys, temples, tourists, and ethnoprimatology. Cultur Anthropol 25:600–624
Fuentes A, Hockings K (2010) The ethnoprimatological approach in primatology. Am J Primatol 72:841–847
Fuentes A (2012) Ethnoprimatology and the anthropology of the human-primate interface Annu. Rev. Anthropol. 41:101–17 doi: 10.1146/annurev-anthro-092611-145808
Fuentes A, Southern M, Suaryana KG (2005) Monkey forests and human landscapes: is extensive sympatry sustainable for *Homo sapiens* and *Macaca fascicularis* in Bali? In: Paterson JD, Wallis J (eds) Commensalism and conflict: the primate-human interface. American Society of Primatologists, Norman, pp 168–195
Gumert MD (2011) The common monkey of Southeast Asia: long-tailed macaque populations, ethnophoresy, and their occurrence in human environments. In: Fuentes A, Jones-Engel L (eds) Monkeys on the edge: ecology and management of long-tailed macaques and their interface with humans. Cambridge University Press, Cambridge, pp 3–45
Hockings K, Humle T (2009) Best practice guidelines for the prevention and mitigation of conflict between humans and great apes. IUCN/SSC Primate Specialist Group, Gland
Hockings K, Yamakoshi G, Kabasawa A, Matsuzawa T (2010) Attacks on local persons by chimpanzees in Bossou, Republic of Guinea: long-term perspectives. Am J Primatol 72:887–896
Janson HW (1952) Apes and ape lore in the middle ages and the renaissance. Studies of the Warburg Institute,Vol. 20. London: Univ. London
Jones-Engel L, Engel GA, Schillaci MA, Babo R, Froelich J (2001) Detection of antibodies to selected human pathogens among wild and pet macaques (*Macaca tonkeana*) in Sulawesi. Am J Primatol 54:171–178
Jones-Engel L, May CC, Engel GA, Steinkraus KA, Schillaci MA, Fuentes A (2008) Diverse contexts of zoonotic transmission of simian foamy viruses in Asia. Emerg Infect Dis 14:1200–1208
Kirksey ES, Helmreich S (2010) The emergence of multispecies ethnography. Cultur Anthropol 255:545–576
Knight J (2006) Monkey mountain as a megazoo: analyzing the naturalistic claims of "Wild Monkey Parks" in Japan. Soc Anim 14:245–264
Kohn E (2007) How dogs dream: amazonian natures and the politics of transspecies engagement. Am Ethnol 34:3–24
Lane KE, Lute M, Rompis A, Wandia IN, Arta Putra IGA, Hollocher H, Fuentes A (2010) Pests, pestilence, and people: the long-tailed macaque and its role in the cultural complexities of Bali. In: Gursky-Doyen S, Supriatna J (eds) Indonesian primates; developments in primatology: progress and prospects. Springer Science, Berlin, pp 235–248
Malone NM, Fuentes A, Purnama AR, Adi Putra IMW (2003) Displaced hylobatids: cultural and economic aspects of the primate trade in Java and Bali, Indonesia. Trop Biodivers 8:41–49
Nijman V (2009) An assessment of trade in gibbons and orang-utans in Sumatra, Indonesia. TRAFFIC Southeast Asia, Petaling Jaya, Indonesia
Ohnuki-Tierney E (1987) The monkey as mirror. Princeton University Press, Princeton
Ohnuki-Tierney E (1995) Representations of the monkey (Saru) in Japanese culture. In: Corby R, Theunissen B (eds) Evaluative proceedings of the symposium ape, man, apeman: changing views since 1600. Leiden University, Leiden, pp 297–308

Palmer M, Bernhardt E, Chornesky E, Collins S, Dobson A, Duke C, Gold B, Jacobson R, Kingsland S, Kranz R, Mappin M, Martinez ML, Micheli F, Morse J, Pace M, Pascual M, Palumbi S, Reichman OJ, Simons A, Townsend A, Turner M (2004) Ecology for a crowded planet. Science 304:1251–1252

Paterson JD, Wallis J (2005) Commensalism and conflict: the primate–human interface. American Society of Primatologists, Lawrence

Riley EP (2007) The human-macaque interface: conservation implications of current and future overlap and conflict in Lore Lindu National Park, Sulawesi, Indonesia. Am Anthropol 109:473–484

Riley E, Fuentes A (2011) Conserving social-ecological systems in Indonesia: human-nonhuman primate interconnections in Bali and Sulawesi. Am J Primatol 73:62–74

Riley E, Priston N (2010) Macaques in farms and folklore: exploring the human–nonhuman primate interface in Sulawesi, Indonesia. Am J Primatol 72:848–854

Riley EP, Wolfe LD, Fuentes A (2011) Ethnoprimatology: contextualizing human/primate interactions. In: Campbell C, Fuentes A, MacKinnon K, Bearder S, Stumpf R (eds) Primates in perspective. Oxford University Press, Oxford, pp 676–686

Rose D (2009) Introduction: writing in the anthropocene. Austr Humanit Rev 49:87

Shepherd CR (2010) Illegal primate trade in Indonesia exemplified by surveys carried out over a decade in North Sumatra. Endanger Species Res 11:201–205

Sponsel LE, Ruttanadakul N, Natadecha-Sponsel P (2002) Monkey business? The conservation implications of macaque ethnoprimatology in southern Thailand. In: Fuentes A, Wolfe LD (eds) Primates face to face: the conservation implications of human-nonhuman primate interconnections. Cambridge University Press, Cambridge, pp 228–309

Sprague DS (2002) Monkey in the backyard: encroaching wildlife in rural communities in Japan. In: Fuentes A, Wolfe LD (eds) Primates face to face: the conservation implications of human-nonhuman primate interconnections. Cambridge University Press, Cambridge, pp 254–272

Wallis J, Lee DR (1999) Primate conservation: the prevention of disease transmission. Int J Primatol 20:803–825

Wong CL (2000) Population dynamics of the feral macaques in the Kowloon Hills of Hong Kong. Am J Primatol 50:53–66

Wheatley BP (1999) The sacred monkeys of bali. Prospect Heights, NY: Waveland

Part IV
Current Scenarios of Human-Macaque Conflict

The exam is History of Mankind.
I stammer and hedge.

One monkey stares and listens with mocking disdain,
the other seems to be dreaming away —
but when it's clear I don't know what to say
he prompts me with a gentle
clinking of his chain.

Wisława Szymborska[1] *1957*

[1] Szymborska W (1998) Poems new and collected [Translated from Polish by Stanislaw Barańczak and Clare Cavanagh]. Harcourt, Orlando

Chapter 8
Gaadli

K.P. Poornachandra Tejaswi
Translated by Honnavalli N. Kumara and Shanthala Kumar

K P Poornachandra Tejaswi (1938–2007) was a remarkable figure in Kannada[1] *literature. Poet, novelist, travel writer, naturalist, and essayist, Tejaswi is credited with ushering in a new realm of "protest literature" with his short-story collection* ***Abachoorina Post Offisu****. Recipient of many national and state awards for his literary contributions, Tejaswi was also deeply interested in science and his science writings charmingly balance science and art. Nature and biological phenomena remained a deep source of inspiration for him throughout his life—much of his work explores the multiple facets of human engagement with the environment and wildlife. In 1962, Tejaswi moved to Mudigere, in the Western Ghats mountain range of southern India, to take up farming and remained there for the rest of his life. His collection of short stories* ***Parisarada Kathe*** *(Stories of the Environment), written during the following decade, is based on his early experiences of living and farming on his coffee and cardamom estate in Mudigere. "****Gaadli****" from* ***Parisarada Kathe*** *is a delightful comment on the complex relationship between people and macaques in this part of the country. Through the trials and tribulations of Gaadli, the eponymous protagonist of the story, the author paints a deeply perceptive picture of the bond that connects people to monkeys. In a succinct piece of writing, Gaadli transcribes the entire gamut of issues surrounding the human-macaque conflict situation; Tejaswi's perspective*

[1] Kannada is one of the many languages of India and is largely spoken by the people of the state of Karnataka in southern India.

H.N. Kumara (✉)
Conservation Biology, Sálim Ali Centre for Ornithology and Natural History, Anaikatty P.O.
Coimbatore 641108, Tamil Nadu, India

National Institute of Advanced Studies, IISc Campus, Bengaluru 560012, India
e-mail: honnavallik@gmail.com

S. Kumar
No. 107, 2nd Floor, Riverstone Ruby Apartment, Vallalar Nagar, Near Chinmaya Vidyalay,
Vadavalli, Coimbatore 641041, Tamil Nadu, India
e-mail: shaanyk@gmail.com

S. Radhakrishna et al. (eds.), *The Macaque Connection: Cooperation and Conflict between Humans and Macaques*, Developments in Primatology: Progress and Prospects 43,
DOI 10.1007/978-1-4614-3967-7_8,

on monkey worship, crop raiding, culling, food provisioning, and translocation will surely strike a chord in primatologists and conservation biologists alike.

The residents of Nidagodu, Hoishalalu, and other villages had called for a meeting; I too attended it as I thought it was to discuss the menace caused by monkeys. Indeed, the agenda was this.[2] But the minute the meeting commenced, all the gathered members began to berate and threatened to beat Gaadli, who worked as a *writer*[3] on Sasi Estate. They were angry because Gaadli had chased the monkeys from his estate, which had then moved into their villages and were creating problems on their estates. If I was in Gaadli's position, I would have said, "if the presence of monkeys on my plantation is a nuisance, I will chase them out of my estate in any direction, and who are you to ask?" But the soft-natured Gaadli, without protesting, contritely said that somebody had brought the monkeys and left them there, and that the *Panchayat*[4] should arrange to catch and release them somewhere else.

Organising for a monkey catcher and capturing the monkeys and releasing them elsewhere is a rather regular business here. If the *Panchayat* sanctioned the money, they would contract monkey catchers from Belur to trap the monkeys and release them elsewhere. Then, the monkeys would start to harass the people there. This had happened once earlier, even before I bought my estate.

People feared killing monkeys as they believed that monkeys are an incarnation of the god Hanuman. If due to any reason a monkey died, people would follow the ritual of treating it with milk and ghee, cremate it, and all the villagers would be in mourning. As monkeys have hands, legs, and faces like humans, people are probably afraid of killing them! But, as monkeys are eaten in the districts of Dakshina Kannada and Kodagu, these monkey-related problems must not exist there. The belief in the divinity of monkeys is so widespread however that even people who belong to the Billava and Harijana castes and whose traditions permit them to eat monkey flesh have been proclaiming that monkeys should not be eaten. There are two Billava families in my village, and when I questioned them about the practice of monkey eating, they behaved as if I had insulted their status and protested loudly. Some others picked up this story and made a song and dance about Billava ancestors who had consumed monkey meat for its medicinal value and castigated the people who had mentioned this to me.

The monkeys here have taken advantage of the religious beliefs of the villagers and damage their property with impunity. I don't think any of the other wild animals cause half as much damage as these monkeys do. As they have highly developed hands and good brains to think with, it is very difficult to control them. When they

[2]The unusual syntax, phrases, and colloquialisms used in this story are native Indian usages. We have deliberately used Indian English in this translation as we feel it more accurately captures the flavour of the original story. For an overview of Indian English, see Sailaja P (2009) Indian English. Edinburgh University Press, Edinburgh

[3]A clerk who is responsible for preparing the administrative documents of a government department or a private institution. In this case, Gaadli's duties included those of an estate manager and an estate dogsbody.

[4]A village council that performs certain administrative and judicial functions.

raid estates, one vigilant animal usually sits atop a tree to watch for people while the rest of them feed on the crop. If anybody approaches, the animal on guard signals the rest by calling *kirro kirro*. Immediately, all the other monkeys escape into the nearby forest. Climbing the tall trees in the forest, they sit behind the foliage and even *Brahma*[5] cannot locate them!

When the *Panchayat* members had chided Gaadli, I had been surprised, as I was not aware of the intensity of the problem created by the monkeys. But within a few days of the meeting, my estate watcher, Maara, came to me saying that he wanted to resign, as it was not possible to control the havoc caused by the monkeys. I went to check on the cardamom plot—it was a disaster! How monkeys can destroy cardamom plants can only be known to people who have seen it. Every cardamom plant had been uprooted, ripped apart, and hung on various trees. I was distressed and furious at the same time.

This group had absolutely no fear of humans. Maara used to beat the drums, burst firecrackers, and do a lot of other things. They were just not bothered by any of these. At the end of it, when Maara went to chase them away, they flashed their big teeth and scared him with their *gurrr-gurrr*. Earlier, Maara had claimed that Hanuman should not be killed, but now that they had frightened him, he alarmed me by declaring that, "If these are not killed with gun, they are a threat to my life and I will resign."

I was not such an *ahimsavadi*[6] to have the monkeys transported to a distant place by paying 10–20 rupees per monkey. And so by the time Maara had decided that the monkeys should be killed, I had also reached a similar state of mind.

My anger against these monkeys had now reached its zenith and the incident in which they attacked Kivi was also a reason. Assuming that on seeing a dog the monkeys would not climb down to the ground, Maara instigated Kivi against them. On seeing the dog, the monkeys vanished into the forests yelling out *taarooo*. Gaining confidence, Kivi chased them into the forest. Having distanced themselves from humans and seeing that the dog was alone, seven to eight large monkeys jumped on the dog and bit him, tearing his flesh out in two or three spots. If Maara had not gone on time and chased them away with a stick, they would have killed Kivi. After this incident, Maara was afraid to tackle the monkeys on his own. I not only had to go to the hospital and get Kivi's wounds stitched but had to make the rounds of the hospital for 14 days to get Kivi his antirabies injections.

I had lost my patience. I was not in a position to consider wildlife conservation, Forest Department laws, and nonviolence. Never mind the monkeys; I was so angry that even if people had caused the same damage, I would have killed them.

Now I understood why people had tried to beat up Gaadli in the *Panchayat* meeting.

---

[5]The god of creation in Hindu mythology.

[6]Literally, "practitioner of *ahimsa* or nonviolence". Although this concept is an essential tenet of Hinduism, Jainism, and Buddhism and therefore, an ancient one, the term itself was popularised by M. K. "Mahatma" Gandhi, the political leader of pre-independent India, who successfully promoted its use as a tool in all spheres of life, including politics.

I loaded my gun and on seeing Kivi with me, one of the monkeys started down the tree, flashing his canines and threatening *gurr-gurr*. The face of the monkey was like the red face of a European, with flashes of anger in its eyes. Maara cried out "Put a bullet to that one, it is the leader of the group!" I fired and the monkey fell instantly to the ground and died.

I had gone with a decision to kill all the monkeys. But when the leader of the group fell from the tree and died, the impact of this on the entire group was terrible. All the monkeys fled helter-skelter, calling out *kurro marro*, and jumping from branch to branch. Maara, in a victorious mood, called out, "Shoot a few more." But there was no chance of that as the monkeys had crossed the paddy field, the plantation, and had fled in the direction of Gaadli's Sasi Estate.

I expected some objections from the villagers for killing the monkey. It seems some people did object. Most of the villagers however expressed their support for killing the monkey and for chasing the rest of the group towards Gaadli's plantation. Hunter Manjappa even threatened Gaadli, "Next time you chase the monkeys in this direction, we will not kill the monkeys and be sinners, we will kill you."

****

The owner of the estate, where Gaadli worked, was Sherav. He spent very little time on the estate as his children worked in Mumbai. The estate appeared derelict and shabby due to lack of maintenance. As the trees had not been pruned regularly, they had grown tall and the canopy was so thick that sunlight did not penetrate to the ground, making it difficult to even chase the monkeys away. Sherav did not spend any money on estate maintenance; instead, he appeared to think that if Gaadli was reprimanded, the estate would flourish. He used to admonish Gaadli saying, "You have ruined the estate by attending to it only once a month." Such rebukes increased severalfold after the monkeys had settled in the estate.

Gaadli tried whatever he could to stop the monkeys. He set a roller in the stream, where it rotated based on the flow of water, and tied it to a metal stick. The stick was placed next to a metal drum, so that the stick continuously beat the drum, creating a load noise. However, the continuous sound coming from this drum had become a lullaby to these monkeys, and they started ignoring it.

He also experimented with a strange method to chase the monkeys—he stored country liquor in a few pots and laid out a few sticks next to the pots. I don't know who suggested this idea to him. His reasoning was that once the monkeys get high after drinking the liquor, they would fight one another with the sticks. When I joked about this method to a few people in the village, they responded quite seriously that it was a regular practice in the region. Taken aback, I enquired with Gaadli about the success of the technique. On one occasion, he said, some people drank up half the liquor and added water to the pots, and the monkeys did not touch the contaminated liquor. Another time the liquor pots were apparently broken but the sticks had not been disturbed; the monkeys had probably fought without using the sticks and had broken the pots.

Sometime later, Gaadli came to see me. The liquor pot experiment had failed. The monkeys, instead of fighting amongst themselves, had started breaking the liquor pots. His estate owner had warned him that he would fire him if he failed to stop the monkey menace. Gaadli sadly asked me to help him by shooting the monkeys.

I asked him, “Why I should come and kill the monkeys? Don’t you know how to shoot?”

“I know to use the gun. But I don’t shoot monkeys, Sir.”

“Why won’t you shoot them?”

“Is not the monkey a god? So far I haven’t killed any monkeys, Sir. How can I kill them now?”

“So what if you haven’t killed the monkeys so far, you kill them now.”

“That is my belief, Sir. Why should I go to hell by killing these monkeys? Tomorrow, if something happens to me…”

So I said, “Okay Gaadli, if you believe that monkeys are gods, you should convince me to not kill them. How can you ask me to hunt them? What if something happens to me after killing the monkeys?”

Gaadli understood and approved my logic. “That is also true, Sir. Why should I make you kill the monkeys and accrue the sin for that?”

I did not want to chase or kill the monkeys in his estate because if monkeys were chased away from there, they would flee to my estate. And as I had seen one *Panchayat* meeting, I was not ready to arrange another one.

Gaadli went to Belur to enquire about the monkey catchers. But as they asked 25 rupees for each monkey, he was unable to pay them and hire their services. Ultimately, Gaadli begged and convinced them to provide a cage on a rental basis and brought the cage back with him by bus.

So far, Gaadli’s monkey-chasing methods had provided much amusement, but few results. Hence, I didn’t ask him about the result of this exercise although I had seen him taking the monkey-catching cage on his bullock cart from the bus stand to his estate.

The monkey-catching cage was a modified version of the cages that are used to transport tigers or lions by circus companies. The cage had two chambers. Both the chambers had doors that could be lifted from the top. The outer door could be lifted up and latched. A wire connected the latch to a rod inside the chamber. If monkeys shook or disturbed the rod inside the chamber, this would trigger the latch and the door would fall and lock the cage. It was not possible to open the door from the inside, even if the monkeys pulled the latch by putting their hands out. It was only possible to lift the door from outside. Once the monkeys were trapped in the first chamber, the door of the second chamber would be lifted and the monkeys sent in here. Then, the outer door would be lifted and latched to trap the next monkey. After all the monkeys were caught in this fashion, they would be taken away in the cage. But I had my doubts about its efficacy. Monkeys are such intelligent creatures, I thought; catching them in this big cage that looks like a jail will be difficult. Even if one or two monkeys do get caught by chance, the others would be frightened and run away.

After a few days, cattle herder Rama told me that Gaadli had caught one or two monkeys in the cage. As Gaadli was not around, I could not talk to him. And as I was busy with estate work, I could not go to meet Gaadli, though I wanted see how he had caught the monkeys.

After 2 days, Rama told me that Gaadli had caught many monkeys and the cage was filled with them.

I was now excited to see the cage. I went to the estate and asked for Gaadli, but by then he had taken the cage on Poodiya's cart to the forest where he would release the monkeys the next morning.

Later, I didn't see Gaadli for a few days. Then, I met two villagers coming down the road who were talking about him. From their talk, it appeared that when Gaadli tried to release the monkeys in the forest, the monkeys came out of the cage and pushed Gaadli inside the cage and locked him in! They told me that Gaadli was inside the cage for 2 days without food. Then, people from a neighbouring village saw him and released him. By this time, he was half dead due to thirst and hunger. I had seen many strange and peculiar things after I came to this forest to tend my plantation. But I could not believe that monkeys had put Gaadli inside a cage and locked him in. Seeing my disbelief, the two villagers said, "If you do not believe our words, ask Gaadli", and they left.

I thought the owners of the nearby estate must have caught Gaadli and put him inside the cage when he went to release the monkeys. Where stories about Gaadli were concerned, people commonly added a hundred things to one.

When I met Gaadli, I asked him, "Is it true that the monkeys locked you inside the cage?" He replied that it was true that he had been trapped inside the cage. But he had neither been pushed inside by monkeys nor by man. He had been trapped inside the cage due to some unexpected events. One can say that the reason for this is the worshipful regard he had for monkeys.

When each monkey was trapped and was made to go to the second chamber, Gaadli thought that they should not die because of hunger. So he provided the monkeys with jackfruit, rice, and water on all the days. When the cage became full, he had taken it on Poodiya's cart to release the monkeys in the forests of Jennapura, about 10 miles away. When he took down the cage from the cart with Poodiya's help and opened the door, only a few monkeys ran out. The monkeys had grandly feasted for many days in the cage and most of them had decided to remain there. Thus, they protested when they were forced out. Poodiya and Gaadli chased out a few monkeys with much effort using sticks, but when they lifted the outer door to free the remaining ones, the monkeys that had been chased out earlier, now went back inside the cage. Poodiya had no love for monkeys. When he tried to chase them out by beating them, he broke a hand of one of the monkeys. This upset Gaadli and he scolded Poodiya. He then suggested that they wait for some time. If the monkeys did not get any food, they would leave the cage to look for some.

By this time, Poodiya was annoyed with the monkeys. When he realised that while waiting for the monkeys to get hungry, they would also go hungry, he angrily said, "I can't wait; you let me know once all the monkeys are out of the cage." He then returned to the village.

As time passed, all the monkeys, one by one, moved out of the cage except for one that remained in the cage, when evening fell, expecting food from Gaadli. Thirsty and hungry, Gaadli had lost all patience; he went inside the cage and pulled the tail of the monkey and pushed it towards the door. The monkey, which had been sitting inside occasionally threatening him by flashing its canines, now screeched and scratched him. Gaadli remembers that this startled him into shaking the metal rod inside the cage. The monkey fled the cage but the door slid down and locked Gaadli inside. Although he tried to lift the door, he could not open it from inside the cage.

Gaadli and Poodiya had come to a remote area in the forest to release the monkeys to prevent them from returning to the estate. There were no people around who could hear Gaadli's cries for help. The nearest village, Manale, with a few houses was about 2 miles away. Gaadli was inside the cage for 2 days without food and water. He was finally in such a state that when he saw two women who had come to collect firewood, he just lay there blinking his eyes, unable even to call for help. The women were so terrified when they saw a human being inside the cage that they threw down the firewood and ran back to the village. On hearing this from the women, some men came from the village to release Gaadli; he just lay there gasping for water.

Gaadli told me that if Poodiya had remained with him, this would not have happened, and he upbraided Poodiya in front of me for the difficult situation he had caused. As Poodiya had not imagined that Gaadli would get caught in the cage, his failure to return, even after 2 days, had not worried him. When I spoke to Poodiya, he retorted "How would I know that he would be in a situation where he would not find food or water when there are houses here, every few steps?"

Acknowledgments We thank Sindhu Radhakrishna and Anindya Sinha for giving us an opportunity to contribute to this volume. We are very grateful to K. Santhosh for his help with the translation, to Sriram of Pustaka Prakashana for help with obtaining copyright permission for the translation, and to an anonymous reviewer for useful comments that helped improve an earlier version of this translation. HNK thanks P. A. Azzez from SACON for his consistent support in all his endeavours.

Chapter 9
Macaque–Human Interactions in Past and Present-Day Sri Lanka

Charmalie A.D. Nahallage and Michael A. Huffman

9.1 Introduction

In spite of Sri Lanka's relatively small size (64,740 km^2), the island supports a high level of biodiversity and endemism (Gunathilleke and Gunathilleke 1983; Erdelen 1988). Part of the reason for this high degree of biodiversity is the variety of habitats found between sea level and the highest peak, 2,524 m asl in the central highlands. The habitat types include mangrove forest, grassland, semiarid thorn forest and shrub land, tropical evergreen rain forests, dry mix evergreen, intermediate moist evergreen, highland evergreen, and temperate forests (Erdelen 1988). There are five primate species in Sri Lanka, the toque macaque (*Macaca sinica*), the gray-handed crested langur (*Semnopithecus priam thersites*), the purple-faced leaf langur (*Trachypithecus vetulus*), and the two nocturnal lorises (*L. tardigradus* and *L. lydekkerianus*). With the exception of *S. priam* and *L. lydekkerianus*, all are endemic and classified as endangered or critically endangered (Dela 2007; Rudran 2007; Nahallage and Huffman 2008).

Sri Lanka is primarily an agricultural country, and over the last few decades, due to agricultural, irrigational, and industrial projects and an increase in the human population and urban expansion, natural forested areas have declined rapidly (Erdelen 1988; Wickramagamage 1998). Owing to fragmentation of forests in the wet and dry zones, primates are increasingly frequenting farms and agricultural plots in search of alternative food resources (Nahallage and Huffman 2008). This is the main cause for primate–human conflict today.

C.A.D. Nahallage (✉)
Department of Sociology and Anthropology, University of Sri Jayewardenepura, Gangodawila, Nugegoda, Colombo, Sri Lanka
e-mail: charmalie2@hotmail.com

M.A. Huffman
Department of Social Systems Evolution, Primate Research Institute, Kyoto University, 41-2 Kanrin, Inuyama, Aichi 484-8506, Japan

S. Radhakrishna et al. (eds.), *The Macaque Connection: Cooperation and Conflict between Humans and Macaques*, Developments in Primatology: Progress and Prospects 43, DOI 10.1007/978-1-4614-3967-7_9, © Springer Science+Business Media, LLC 2013

Fig. 9.1 People and primates coexist harmoniously in religious temple grounds. Religious devotees pray and provide food for monkeys as religious offering at the Ruwanwelisaya Stupa (*upper left, right, lower left*). A priest resting on the steps with gray langurs grooming nearby in the Kataragama temple grounds (*lower right*) (Photos by M. A. Huffman)

However, in Sri Lanka, the peoples' religious beliefs and cultural practices play an important role in defining the terms of primate–human interactions and reflect the past, present, and potential future of their coexistence. An important feature of this country is the presence of monkeys at Buddhist and Hindu temples. For example, the North Central Province city of Anuradhapura, formerly the ancient capital of the country from the fourth century BC to the eleventh century AD, has been protected as a holy site. Named as a UNESCO World Heritage Site, Anuradhapura is visited daily by a large number of local and international tourists alike (Fig. 9.1). The expansive area of woodland surrounding this religious site, protected from development due to its cultural and religious significance, is home to a large population of primates that forage on the natural vegetation. The monkeys are habituated to humans and freely range in and around the sacred sites undisturbed by tolerant monks and tourists. The constant supply of food offerings given to them by religious followers and tourists keep macaques and langurs around the temple grounds. Like Anuradhapura, there are many large and small Buddhist and Hindu temples across the country where primates are protected by the peoples' religious and cultural beliefs, i.e., Kandy, Sigiriya, Dambulla (Central Province), Kelaniya, Kalutara (Western Province), Rumassala, Sella Kataragama,

Kataragama, Sithulpawwa, and Vadasitykanda (Southern Province). These are places where primate–human interactions are relatively harmonious (Fig. 9.1).

Among the five primate species inhabiting Sri Lanka, more is known about the behavioral ecology of the toque macaque than the other primate species. An impressive longitudinal study begun in the 1970s investigated various aspects, such as the social behavior, demography, ecology, disease etiology, and conservation of the toque macaque population in Polonnaruwa (another protected religious-historical site in the North Central Province). The results from this long-term project (e.g., Dittus 1977, 1986, 1998, 2004; Hoelzer et al. 1994; Ekanayake et al. 2007) form the majority of detailed published information about this endemic macaque species today.

One area of research that has received little attention in Sri Lanka is the growing field of ethnoprimatology, the multidisciplinary investigation of humans and nonhuman primate interactions (e.g., Loudon et al. 2006; Paterson and Wallis 2005; Wolfe and Fuentes 2007; Fuentes and Hawkins 2010; Hill and Webber 2010). In this chapter, we present results from questionnaires and field surveys investigating how Sri Lankans view primates, their cultural practices, beliefs, and the state of primate–human interactions in a rapidly developing and changing country. It is not the goal of this chapter to provide a picture of the conservation status of primates in the country but rather to describe the relationship humans have had with macaques and other primate species in general, from both a historical and contemporary perspective. We address such questions as: What place does the toque macaque occupy in Sri Lankan folklore and literature? How do people relate to the toque macaque vis-à-vis the langurs in Sri Lanka? What are the major causes of conflict between toque macaques and humans in Sri Lanka?

9.2 Methods

The study was conducted by direct observation and through formal interviews using a standardized questionnaire. Informed consent was obtained before each interview. The study abided by all laws of Sri Lanka, and the protocol and permission to conduct the study was approved by the Department of Wildlife Conservation. A total of 307 formal interviews were conducted (129 women, 178 men) with participants ranging in age from 18 to 85 years. A breakdown of the respondents by status is listed in Table 9.1. The data presented here were collected during our field visits (5–20 February, 2007; 23 February–3 March, 2009, $N=127$) and by trained undergraduate students from the University of Sri Jayewardenepura and the University of Uva during the period 2007–2009 ($N=180$).

Administratively, Sri Lanka is divided into 9 provinces and 25 districts (Fig. 9.2). Students were selected based on their province and area of origin to obtain data from as many different districts as possible. In total, 23 districts from these 9 provinces are represented in this database. A breakdown of the questionnaires by province is given in Table 9.2. The unbalanced representation of the provinces is

Table 9.1 Breakdown of the status of respondents

Respondent status	Frequency
Farmer	48
Small-scale vegetable and fruit vendor	26
Student	84
People in national parks	26
Housewife	19
Retired government officer	14
Teacher	7
Laborer	6
Buddhist priest	6
Private business owner	4
Traditional doctor	2
Security guard	2
Not specified	25

due to the war in the Northern and Eastern Provinces during the above study periods, which made it difficult to visit or to find students originating from these areas to conduct interviews.

Respondents were shown close-up, full-body pictures of each primate species with their common names in Sinhalese, Tamil, and English to minimize confusion and increase accuracy. The questionnaire included 28 questions on topics including primates species seen in the area, the approximate number of groups and their group size, (whether they damage crops) their preferred food items, the measures taken by the people to prevent primates damaging the crops, whether they were aware of hunting of primates for food in their respective areas, use of primate body parts for medicinal and ritual purposes, and primate myths and folklore known of or heard about by the respondents.

9.3 Results

9.3.1 *The Perceived Trends of Local Primate Population Size and Damage to Crops*

When asked about the current number of primates in their area compared to earlier times, 51% of all respondents believed that they had increased in recent years. (It was not our intention to use this response as a measure of population size but rather as a means of estimating any relative change in the frequency of contact between people and primates in recent times.) In some provinces, the majority of respondents said that contact with primates was increasing (Central, 92%; North Central, 75%; Uva, 71%; North Western, 68%; Southern, 53%), while in other provinces, the majority thought their numbers were declining (Western, 41%; Eastern, 64%; Sabaragamuwa, 59%). Eighty-nine percent of respondents said that primates raid

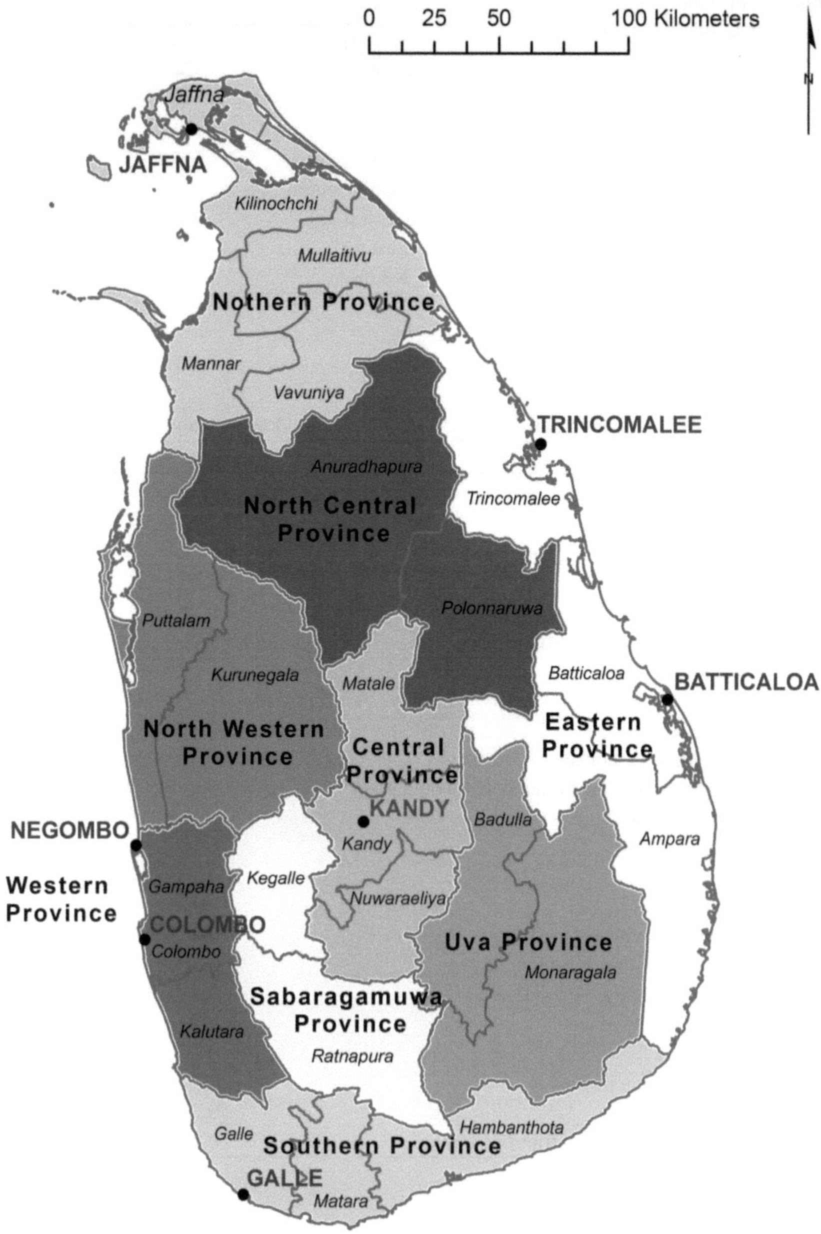

Fig. 9.2 Administrative province and district map of Sri Lanka (Source: Produced by Dr. Janet Nackoney)

Table 9.2 Province-wise distribution of questionnaires

Province	Percentage respondents
Western	26
Southern	24
Sabaragamuwa	15
North Western	10
Uva	7
North Central	7
Central	5
Eastern	5
Northern	3

Table 9.3 Prevention methods used against crop-raiding monkeys ($N=307$)

Method	Percentage usage
Firecrackers	31
Making loud sounds	22
Shooting	10
Using traps	9
Throwing stones	7
Catapulting/sling-shooting stones	5
Scarecrow	3
Using dogs	2
Poisoning	2
Hanging dead monkeys in the trees	1
Placing cut trees and branches around garden	1
Covering crops with nets	1
Other methods	6
Using mirrors	0.8
Guarding	0.8
Hanging polythene strips, bags	0.6
Applying cow dung	0.6
Hanging red flags and umbrellas	0.6
Hanging shiny objects	0.5
Using nets	0.3
Sprinkling gun powder	0.3
Dynamite	0.2
Applying black oil to fruit tree trunks	0.2

crops, and 60% considered them as pests. Forty-eight percent thought that primates were afraid of people, while 45% said they were not, and some (7%) had no opinion one way or the other.

People used various techniques to protect their crops from monkeys (Table 9.3). Respondents on average reported 2.5 (SD 1.38, range 1–6) different methods; the

Fig. 9.3 Agriculture pests, temple guardians, or entertainers, macaques are regarded in many different ways across the country. Small-scale farmers selling their produce at roadside stands near their homes (*upper left*), commercial farmers (*upper right*), and roadside food vendors (*lower left*) are often the target of crop-raiding monkeys (Photos by M. A. Huffman). A toque macaque performing in front of a private home in a suburb near Colombo (*lower right*). (Photo by C. A. D. Nahallage)

most common was the use of firecrackers (31%) or making other loud sounds (22%). Trapping or shooting monkeys (19%) was practiced most frequently in the Southern, Sabaragamuwa, and North Western Provinces.

9.3.2 Use of Nonhuman Primates as Pets and Performers

Most people said they preferred toque macaques as pets because they believed that macaques resemble humans more closely than langurs and adapt better to captivity. In North Central, North Western, and Sabaragamuwa Provinces, toque macaques are used as performers for profit (Fig. 9.3). Trained macaques are taken to urban neighborhoods, villages, and to public places such as parks and temple grounds to entertain the crowds. The most common performances were scenes such as visiting parents-in-law during the New Year, carrying a box of gifts on the head while walking upright on hind legs, or being punished by the police for stealing. Toque individuals are usually dressed in a pair of trousers and a shirt and constrained by a leash. Typically, the cost is equivalent to one US dollar per performance.

Table 9.4 Use of primate body parts for medicinal purposes

Species/part	Ailment	C	N	NC	NW	E	Sb	S	U	W
Langur meat	Asthma	○		○		○	○		○	○
Meat	Poor eyesight						○	○		○
Meat	Leprosy					○				
Meat	Malnutrition, piles			○						
Meat	Malnutrition							○	○	
Meat	Heartburn, kidney and lung diseases						○		○	
Meat	Boils, TB									○
Meat	Heart and lung diseases							○	○	
Liver	Malnutrition			○				○		
Heart	Malnutrition							○		
Hands, tail	Cracks in hands and feet								○	
Stomach contents	Malnutrition			○						
Oil	Burns									○
Macaque meat	Piles				○					
Meat	Asthma				○					
Meat	Whooping cough, piles	○								
Meat stomach content	Malnutrition	○								
Feces	Whooping cough				○					
Oil	Fractures							○		○
Urine	Snake bites							○		
Loris tears	Not specified			○						

Province: *C* Central, *N* Northern, *NC* North Central, *NW* North Western, *E* Eastern, *Sb* Sabaragamuwa, *S* Southern, *U* Uva, *W* Western

9.3.3 Primates as Food, Medicine, and Ritual Objects

With the exception of Northern Province, primates are hunted for food everywhere but at a very low frequency and in secret. Forty-seven percent noted that people in their respective areas hunted primates for food, while 42% said hunting did not occur, and 11% said they were not sure.

Primates are also hunted for body parts to be used as medicine (Table 9.4) and for ritual activities; this occurred very infrequently and only in rural areas. There are marked province-wise similarities and differences in this belief system, suggesting that it was historically practiced throughout the country. In Central Province, for "apala" (ritual practiced against unfavorable positions of the planets in one's horoscope), the intestine of a macaque is wrapped around the neck of an affected person at dawn, and he/she is not allowed to speak during the treatment. In North Western Province, loris tears mixed with other ingredients are ritually smeared on a plate to see into the future. In Sabaragamuwa Province, the heart of the purple-faced langur is given to pregnant women, and the skull, skin, and penis are used in "thovil," another traditional healing ritual. Also, the meat and bones are used for "huniyam" (a practice akin to black magic or voodoo). In Uva Province, langur heart, when consumed, is believed to increase one's intelligence. In Western Province, both toque macaque and purple-faced langur skins are used to make drums.

9.3.4 Primates in Folklore, Myth, and Other Narrations

Folklore and myths regarding primates are abundant in Sri Lanka. The majority of the population practice Buddhism, introduced approximately 2,500 years ago, and many related folklore regarding primates can be found. The Jataka stories, originating from India, revolve around the 550 rebirths of the bodhisattva before attaining buddhahood. In 22 of these stories, the bodhisattva was born as a monkey, and these stories highlight the kindness, forgiveness, helpfulness, intelligence, and patience of bodhisattva through the behavior of monkeys in different situations (e.g., Nalapana Jataka, Kapi Jataka, Vanarinda Jataka, and Tayodhamma Jataka). Other primate folklore in Sri Lanka also highlight their intelligence and curiosity. For example, the monkey that lost its tail by being too curious or the one that injured its face by trying to imitate a person shaving his beard. One that appears in many cultures is about the hat seller who got his hats back from a troop of thieving monkeys by getting the troop to imitate his actions, taking off his own hat, and throwing it to the ground. A famous Hindu epic poem, the *Ramayana*, is also very popular in Sri Lanka with both the Hindus and the Buddhists. A central character, Hanuman (the monkey god worshipped by Hindus), is the brave leader of an army of monkeys, who flies to Sri Lanka in search of Sita, kidnapped by the Sri Lankan king, Ravana (Fig. 9.2).

Myths and beliefs about primates differ across the country. For example, some communities in the Northern and Western Provinces believe that it is good luck to see monkeys when they leave the house, whereas some communities in North Western, North Central, Southern, and Western Provinces believe that if a monkey calls out just before someone leaves the house for work in the morning, it will bring bad luck and the person will not be able to carry out their work as planned. In Southern, Uva, and Western Provinces, people believe that the right side of a langur's body is made of human flesh, so they refrain from eating them altogether or do not use any part of the right side of the animal's body. In Sabaragamuwa Province, people believe that if a monkey enters into the house through the kitchen door, somebody in the house will fall seriously ill. It is also believed that macaques were created by the demon "Wasawarthi Mara" to cause trouble for villagers. Another animal created by him is thought to be the wild pig, another source of damage to garden plots.

9.4 Discussion

Like other "weed species" in the genus *Macaca,* in our survey, toque macaques were frequently found in close proximity to human habitations (Guatier and Biquand 1994; Lee and Priston 2005; Richard et al. 1989). The gray langur and purple-faced langur can also be classifiable as a kind of "weed species," given their ability to live close to human habitations. Though their diets are less omnivorous than macaques, their food habits do overlap, and they forage in cropfields and kitchen gardens in villages, towns, and cities across the country. Not surprisingly, this close proximity to humans has long been a source of conflict, which is mirrored by similar situations throughout the distribution of other macaque species and langurs across Asia (e.g., Fuentes et al. 2005; Fuentes and Hawkins 2010; Riley 2007; Sha et al. 2009; Southwick et al. 1961; Sponsel et al. 2002; Watanabe and Muroyama 2005).

Long-term analysis of trends showing an increase in contact and conflict between humans and macaques was documented in Japan by Watanabe and Muroyama (2005). They attributed this increase in the visibility of previously timid macaques to species range expansion caused by deforestation and a resultant decrease in natural food resources, increased acculturation to human environments, and a less fearful attitude towards humans due to the aging population of farmers living in the countryside. Respondents to our questionnaire in some parts of Sri Lanka believed that primate populations had increased in the last several years. In the absence of reliable studies on regional primate densities, it is not possible to affirm or refute these claims. In the absence of reliable primate population estimates for the country, we can only interpret our results as indicative of an increase in contact with primates, possibly due to some of the same factors identified by Watanabe and Muroyama (2005), e.g., forest fragmentation and habitat loss due to an increase in development activities. Interestingly, many people opined that toque macaques and langurs were forced to come closer to human settlements in search of food resulting in increased contact with people. Respondents in Galle District, Southern Province, stated that earlier, purple-faced langurs were seen only in the forests and hardly came to the village, but with the reduction of local forests, they are now frequent visitors. In the more urbanized and industrialized Western Province, respondents said that the purple-faced langur population had decreased because they saw fewer of them in recent years. Rudran (2007) states that only a few forest patches remain at the edge of the province and that most langur populations are restricted to these patches and to the surrounding areas.

Toque macaques and langur species are not distributed equally throughout Sri Lanka (Nahallage and Huffman 2008); as a result, the damage caused by them differs with location. For example, in places where all three species can be found (Matale District and Nuwaraeliya District, Central Province, Badulla and Monaragala Districts, Uva Province), toque macaques are considered the most serious crop raiders. Also in Central and North Western provinces, where toque macaque numbers are said to be larger, they are considered as more serious pests than langurs. Purple-faced langurs are the most prominent species found in Western and some parts of Southern Province (Galle District). They cause more crop damage and are considered greater pests in these areas than toque macaques. In some parts of the Southern (Hambanthota District) and North Central Provinces, gray langurs cause greater crop damage than purple-faced langurs or toque macaques.

The methods used to prevent crop raiding differed from place to place and was largely dependent on the socioeconomic status of the area (Nahallage and Huffman 2008).

In Western Province, the most common methods used to prevent crop raiding were firecrackers, throwing stones, and making loud sounds. These methods were quite effective in chasing purple-faced langurs away and are harmless. Here, purple-faced langurs were mainly seen in home gardens eating food grown for household consumption. For this reason, they were considered more of a nuisance than a serious pest. For the most part, people in this region tolerated them. In Central, North Central, North Western, Southern, and Sabaragamuwa Provinces, areas of large-scale commercial farming, macaques caused more serious economic damage (Fig. 9.3) and were less favorably looked at. In these cases, people had harsher opinions and used stronger methods to prevent macaques from raiding crops, such as shooting and trapping.

Among the indigenous people of lowland South America, primates are viewed as both symbols of power and as food. In a review by Cormier (2006), a general tendency was noted for larger bodied primates to be hunted for food; however, they were also more likely to be associated with taboos and not eaten by individuals of certain ritual (age, sex, reproductive state) or social status or revered as having come from former human beings or created by a divine being. Some of the indigenous communities avoid certain primate species as food because of their close resemblance to humans (Cormier 2006). In our survey, some respondents said that primates were hunted for food, but that it was not a common practice. The main reason for this low frequency is religious as Buddhism and Hinduism eschew killing animals (Nahallage and Huffman 2008; Rudran 2007). When hunted, however, the langur species were reportedly preferred over toque macaques because of their larger body size. In many areas, people believed that langur meat is especially good for treating certain diseases, such as asthma, leprosy, malnutrition, and piles. As langurs are leaf-eating primates, the respondents reasoned that many of the leaves eaten by the langurs are of high medicinal and nutritional value; therefore, the meat should be good for use as medicine. Furthermore, it is interesting to note that most people in the rural villages believed that primate organs, especially heart and lungs, can be transplanted to humans. This belief likely originates from the close similarity they perceive between primates and themselves, not only in anatomy but in social behavior as well. Another important feature of human–primate interaction in Sri Lanka is the presence of a large number of primates in and around the Buddhist and Hindu temples. There are many famous temples throughout the country visited daily by large numbers of pilgrims. These places play an important role in conserving endangered primates. Other examples from different countries include Lopburi in Thailand; Chamundi Hills in Mysore, India; and Ubud Monkey Forest in Bali (Fuentes et al. 2005; Loudon et al. 2006; Watanabe et al. 2007; Wheatley 1999; Wolfe and Fuentes 2007; authors' personal observations).

In habitat-source countries, primates are intricately enmeshed into the daily lives through folktales and myths (i.e., Ashliman 2011; Cormier 2006; Loudon et al. 2006; Riley and Priston 2010; Shahar 1992). People have likely incorporated monkeys so easily into these stories because of the close similarity between the two species. People in the rural areas of Sri Lanka still strongly believe these myths about primates, and the fear of retribution prevents the majority of people from harming monkeys. The major threat for their survival is the loss of habitat and conflict with farmers, which is more pronounced in some provinces than others. Although farmers believe that primates cause more damage to their crops than other animals, studies elsewhere have shown that damage cause by primates is actually far less than what the farmers believe it to be (Riley 2007; Siex and Struhsaker 1999). The large body size and group size of the primates magnify the actual extent of damage caused (Nahallage and Huffman 2008). Therefore, it is necessary to systematically quantify the damage caused by primates and to communicate these findings to farmers.

Peoples' religious beliefs and cultural practices play an important role in defining the terms of primate–human interactions and reflect the past, present, and potential future of their coexistence (Loudon et al. 2006). In this light, the study of ethnoprimatology is an important conservation tool for understanding the human perspective on primates, which, when meshed with scientific studies, offers a holistic understanding of the current plight of primates. A better understanding of the Sri Lankan perception of macaques and other primates, with whom they coinhabit the island, will be helpful in the conservation and management of primates in Sri Lanka.

Acknowledgments The information presented adhered to all national laws of Sri Lanka in which the research was conducted. The authors thank HOPE Project, a core-to-core program sponsored by the Japan Society for the Promotion of Science (JSPS), for travel funds to carry out the 2006, 2007, and 2009 field visits. We are grateful to our friends and colleagues, Nadeera Kuruppu, Tharindi Weerasingha, Nelum Yapa, Sisira Siriwardene, and Prof. Singha Kumara, for their assistance and guidance in the field and in Colombo. We give special thanks to Dr. Janet Nackoney for producing Fig. 9.2.

References

Ashliman DL (2011) Folklore and mythology electronic texts. University of Pittsburg. http://www.pitt.edu/~dash/folktexts.html. Accessed 20 Oct 2011

Cormier L (2006) A preliminary review of neotropical primates in the subsistence and symbolism of indigenous lowland South American peoples. Ecol Environ Anthropol 2:14–32

Dela JDS (2007) Seasonal food use strategies of *Semnopithecus vetulus nestor*, at Panadura and Piliyandala, Sri Lanka. Int J Primatol 28:607–626

Dittus WPJ (1977) The ecology of a semi-evergreen forest community in Sri Lanka. Biotropica 9:268–286

Dittus WPJ (1986) Sex differences in fitness following a group take-over among toque macaques: testing models of social evolution. Behav Ecol Sociobiol 19:257–266

Dittus WPJ (1998) Birth sex ratios in toque macaques and other mammals: integrating the effects of maternal condition and competition. Behav Ecol Sociobiol 44:149–160

Dittus WPJ (2004) Demography: window to social evolution. In: Thierry B, Singh M, Kaumanns W (eds) Macaque societies: a model for the study of social organization. Cambridge University Press, Cambridge, pp 87–116

Ekanayake DK, Welch DM, Kieft R, Hajduk S, Dittus WPJ (2007) Transmission dynamics of *Cryptosporidium* infection in a natural population of non-human primates at Polonnaruwa, Sri Lanka. Am J Trop Med Hyg 77:818–822

Erdelen W (1988) Forest ecosystems and nature conservation in Sri Lanka. Biol Conserv 43:115–135

Fuentes A, Hawkins KJ (2010) The ethnoprimatological perspective. Am J Primatol 72:841–847

Fuentes A, Southern M, Suaryana KG (2005) Monkey forests and human landscapes: is extensive sympatry sustainable for *Homo sapiens* and *Macaca fascicularis* in Bali? In: Paterson JD, Wallis J (eds) Commensalism and conflict: the primate–human interface. American Society of Primatologists, Norman, pp 168–195

Guatier J-P, Biquand S (1994) Primate commensalisms. Revue D'Ecologie (La Terre et la Vie) 49:210–212

Gunathilleke CVS, Gunathilleke IAUN (1983) A forestry case study of the Sinharaja rainforest in Sri Lanka. In: Hamilton LS (ed) Forest and freshwater development and conservation in Asia and the Pacific. Westview Press, Boulder, pp 289–357

Hill CM, Webber AD (2010) Perceptions of non-human primate in human-wildlife conflict scenarios. Am J Primatol 72:919–924

Hoelzer GA, Dittus WPJ, Ashley MV, Melnick DM (1994) The local distribution of highly divergent mitochondrial DNA haplotypes in toque macaques (*Macaca sinica*) at Polonnaruwa, Sri Lanka. Mol Ecol 3:451–458

Lee PC, Priston NEC (2005) Human attitudes to primates: perceptions of pests, conflict and consequences for primate conservation. In: Paterson JD, Wallis J (eds) Commensalism and conflict: the primate-human interface. American Society of Primatologists, Norman, pp 1–23

Loudon JE, Howells ME, Fuentes A (2006) The importance of integrative anthropology: a preliminary investigation employing primatological and cultural anthropological data collection methods in assessing human-monkey co-existence in Bali Indonesia. Ecol Environ Anthropol 2:1–13

Nahallage CAD, Huffman MA (2008) Diurnal primates in Sri Lanka and people's perception of them. Primate Conserv 23:81–88

Paterson JD, Wallis J (2005) Commensalism and conflict: the primate-human interface. American Society of Primatologists, Norman

Richard AF, Goldstein SJ, Dewar RE (1989) Weed macaques: the evolutionary implications of macaque feeding ecology. Int J Primatol 10:569–594

Riley EP (2007) The human–macaque interface: conservation implications of current and future overlap and conflict in Lore Lindu National Park, Sulawesi, Indonesia. Am Anthropol 109: 473–484

Riley EP, Priston NEC (2010) Macaques in farms and folklore: exploring the human-nonhuman primate interface in Sulawesi, Indonesia. Am J Primatol 72:848–854

Rudran R (2007) A survey of Sri Lanka's endangered and endemic western purple-faced langur (*Trachypithecus vetulus nestor*). Primate Conserv 22:139–144

Sha JC, Gumert MD, Lee BP, Jones-Engel L, Chan S, Fuentes A (2009) Macaque–human interactions and the societal perceptions of macaques in Singapore. Am J Primatol 71:1–15

Shahar M (1992) The Lingyin Si Monkey disciples and the origins of Sun Wukong. Harv J Asiat Stud 52:193–224

Siex KS, Struhsaker TT (1999) Colobus monkeys and coconuts: a study of perceived human-wildlife conflicts. J Appl Ecol 36:1009–1020

Southwick CH, Beg MA, Siddiqi MR (1961) A population survey of rhesus monkeys in villages and temples of northern India. Ecology 42:538–547

Sponsel LE, Ruttanadakul N, Natadecha-Sponsel P (2002) Monkey business? The conservation implications of macaque ethnoprimatology in Southern Thailand. In: Fuentes A, Wolfe LD (eds) Primates face to face: the conservation implications of human-nonhuman primate interconnections. Cambridge University Press, Cambridge, pp 288–309

Watanabe K, Muroyama Y (2005) Recent expansion of the range of Japanese macaques and associated management problems. In: Paterson JD, Wallis J (eds) Commensalism and conflict: the primate-human interface. American Society of Primatologists, Norman, pp 400–419

Watanabe K, Urasapon N, Malaivijitnond S (2007) Long-tailed macaques use human hair as dental floss. Am J Primatol 69:940–944

Wheatley BP (1999) The sacred monkeys of Bali. Waveland Press, Prospectus Heights
Wickramagamage P (1998) Large-scale deforestation for plantation agriculture in the hill country of Sri Lanka and its impacts. Hydrol Process 12:2015–2028
Wolfe L, Fuentes A (2007) Ethnoprimatology: contextualizing human and non-human primate interactions. In: Campbell CJ, Fuentes A, Mackinnon KC, Panger M, Bearder S (eds) Primates in perspective. Oxford University Press, Oxford, pp 691–701

Chapter 10
Monyet Yang Dihargai, Monyet Yang Dibenci: The Human-Macaque Interface in Indonesia*

Jeffrey V. Peterson and Erin P. Riley

10.1 Introduction

The worlds of human and nonhuman primates are increasingly overlapping. The term "interface" aptly encapsulates all aspects of the relationships linking together human and nonhuman primates, that is, their dynamic interaction in their shared environment. It is now well understood that there is a long history of sympatry between populations of human and nonhuman primates, one that has experienced many types of changes but most notably in the degree of negative influence from humans. With this knowledge, we can no longer assume that any given nonhuman primate population is absolutely free from human influence. This influence may be most visible in the form of bushmeat markets or anthropogenic habitat conversion but can also be more subtle such as with long-term effects on individual or group level nonhuman primate fitness. Additionally, researchers have become increasingly interested in the epidemiological significance of this interface. Zoonoses co-affecting human and nonhuman primates are continually being researched, providing more detailed information on diseases already familiar to us and occasionally even leading to new zoonoses being discovered (Engel et al. 2006).

While there is considerable variability in the nature of the human-nonhuman primate interface in different areas of the world, a common thread is the concept of space. The overlapping lives of human and nonhuman primates often result from limitations of space and the increasing necessity to share. Much of the human-nonhuman primate interface can be understood by examining how both groups react to sharing space with each other, historically as well as in more recent contexts. Methods in behavioral observation are useful for exploring the reactions of our

*The Revered Monkey, the Despised Monkey

J.V. Peterson (✉) • E.P. Riley
Department of Anthropology, San Diego State University,
San Diego, CA 92182-6040, USA
e-mail: petersonjv@gmail.com

S. Radhakrishna et al. (eds.), *The Macaque Connection: Cooperation and Conflict between Humans and Macaques*, Developments in Primatology: Progress and Prospects 43,
DOI 10.1007/978-1-4614-3967-7_10, © Springer Science+Business Media, LLC 2013

nonhuman primate cousins, while ethnographic techniques focusing on perceptions of local nonhuman primates are proving to be strong indicators of our own attitudes (Lee and Priston 2005).

The purpose of this chapter is to discuss the diverse ecological and cultural facets of the human-macaque interface in Indonesia and explore how these facets intersect to result in a constantly shifting relationship between commensalism and conflict, tolerance and intolerance, and reverence and disdain throughout the country. We begin by describing the high level of macaque species diversity in Indonesia, the ecological contexts in which they are found, and the concomitant cultural diversity of the archipelago. We then explore the ecological (anthropogenic habitat disturbance, crop raiding, hunting, and disease transmission) and cultural (mythology, folklore, and religion) facets of the human-macaque interface. To provide depth to this discussion, we present preliminary findings from a case study of Balinese transmigrant communities in South Sulawesi that aims to understand how Balinese perceptions of macaque sacredness (or reverence) may be impacted by their migration to a new environmental context outside of Bali. We conclude by discussing potential future directions of the human-macaque interface and its relevance for future research and conservation efforts.

10.2 Macaque Species Diversity and Ecological Context

10.2.1 Indonesian Macaques

Macaca is the most geographically widespread nonhuman primate genus and one of the most speciose (Abegg and Thierry 2002). Indonesia is home to ten different macaque species, making it the greatest number of macaque species within a single habitat country (Table 10.1). Of these ten species, *M. fascicularis* is of least concern (IUCN 2010), six are listed as vulnerable (*M. brunnescens, M. hecki, M. nemestrina, M. nigrescens, M. ochreata*, and *M. tonkeana*) (IUCN 2010), one is endangered (*M. maura*) (IUCN 2010), and two are critically endangered (*M. nigra, M. pagensis*) (IUCN 2010).

10.2.2 Macaque Habitat Ecology

There are over 17,000 islands in the Indonesian archipelago that together comprise a wide range of ecological systems. Macaques occur primarily in the western half of Indonesia and extend as far east as the island of Sulawesi. Within this area, the ecological contexts in which macaques are located fall into three main categories: urban/temple environments, protected forest areas, and forest-farm matrix. These three environments, however, are not closed ecological systems. There are many

Table 10.1 Indonesian macaque species

Macaca species	Common name	Geographic location	Habitat[a, b]
1. *M. fascicularis*	Long-tailed	Sumatra, Java, Bali, Kalimantan	UT/PA/FFM
2. *M. pagensis*	Mentawai	Mentawai Islands	FFM
3. *M. nemestrina*	Pig-tailed	Sumatra, Kalimantan	FFM
4. *M. tonkeana*	Tonkean	Central Sulawesi	PA/FFM
5. *M. nigra*	Black crested	Northeastern Sulawesi	PA/FFM
6. *M. nigrescens*	Gorontalo	North Sulawesi	PA/FFM
7. *M. hecki*	Heck's	Northwestern Sulawesi	PA/FFM
8. *M. maura*	Moor	Southwestern Sulawesi	PA/FFM
9. *M. ochreata*	Booted	Southeastern Sulawesi	PA/FFM
10. *M. brunnescens*	Buton	Buton and Muna islands	PA/FFM

[a]*UT* urban/temple environments, *PA* protected areas, *FFM* forest-farm matrix
[b]Source: IUCN (2010)

instances of these ecological contexts neighboring each other or even overlapping. As a result, a single macaque population may find each of these three contexts within its home range.

The urban/temple environmental context is an almost exclusively Balinese phenomenon. The continuous presence of large macaque populations at Hindu temple sites has been well documented (Wheatley 1999; Loudon et al. 2006; Fuentes et al. 2007). These locations constitute urban environments in which Indonesian macaques are able to thrive, though this is not always the case. It has been recently suggested that the long-term close relationship between humans and macaques at the Balinese temple sites may have negative health effects for the macaques such as low-quality diet and bidirectional disease transmission (Lane et al. 2010).

Many of the Indonesian macaques exist in forested habitats that have been designated as protected areas. Given the high levels of biodiversity across the archipelago, protected area conservation is the dominant model of conservation. There are a total of 965 protected areas in Indonesia, representing 12.5% of the total land cover (WRI 2006). Outside of Bali, the existence of protected area habitat may be critical for those species living in areas with high human population densities. For example, the endangered crested black macaque, which occurs in the northeast corner of Sulawesi where human population density is the highest (132 individuals/km^2) (BPS 2000), has been found to maintain stable populations only in areas protected by government mandates, although still in close contact with humans through ecotourism (Melfi 2010).

The forest-farm matrix is the predominant setting for the human-macaque interface in Indonesia (Riley and Priston 2010). The primary issue is that agricultural crops (both subsistence and cash) are planted and cultivated where forest once stood. As a result, these crop gardens typically share a border with macaque forest habitats, thereby simultaneously facilitating foraging access for the macaques and hampering farmers' ability to defend their crops. Crop raiding often causes local farmers to have negative perceptions of nonhuman primates (Lee and Priston 2005). Therefore, the prevalence of the forest-farm matrix may indicate that a large part of the human-macaque interface in Indonesia is characterized by conflict.

10.3 Ethnic and Religious Identity in Indonesia

Within the Indonesian archipelago, there are 6,000 inhabited islands that together hold a population of over 245 million people (CIA 2011). As one of the world's largest populations, Indonesia is also one of the most culturally diverse. The number of distinct ethnic groups has been counted as over 300 (Hoey 2003). Each of these cultures has their own language, cultural traditions, and customary practices, which creates a wide range of ethnicities that comprise Indonesian society. To linguistically unite this diverse archipelago, *Bahasa Indonesia* was made the national language and is taught throughout Indonesia from elementary school onward, along with English. This prioritization of *Bahasa Indonesia* is becoming more and more useful as intercultural contact is increasing in frequency through the ease of intra- and interisland travel and communication.

The Indonesian transmigration program has played a large role in facilitating intercultural contact. This government-sponsored program was funded largely by the World Bank and was intended to alleviate increasing population pressure on the islands of Java, Bali, Lombok, and Madura (Whitten et al. 1987; World Bank 1988). Approximately 334,581 people were relocated between the years 1956 and 1974, well below the goal of roughly three million (Davis 1976). While this program technically ended after the 1980s, unsponsored transmigrants still migrate out of Bali or between transmigrant sites already located on the other islands (World Bank 1988).

Indonesia's cultural diversity is also manifested in people's religious identity. While most Indonesians identify with one of the five officially recognized religions (Islam, Protestant Christianity, Roman Catholicism, Hinduism, and Buddhism (Schiller 1996)), many still adhere to pagan beliefs and traditions (Atkinson 1983). The vast majority of citizens (86.1%) are Islamic, followed by Protestant (5.7%) and Roman Catholic (3%) (CIA 2011). These broad religious categories serve to unite many otherwise very different ethnic groups throughout Indonesia.

10.4 The Human-Macaque Interface: Ecological Facets

10.4.1 Anthropogenic Habitat Alteration

Anthropogenic habitat alteration via such activities as selective logging and forest fragmentation can have a negative impact on nonhuman primate livelihoods, especially when key food species are eliminated within the nonhuman primate group's range. Researchers interested in the potential effect of anthropogenic habitat disturbance on nonhuman primate populations' habitat quality often focus on changes in population size and density (e.g., Bishop et al. 1981; Marshall et al. 2006; Paciulli 2010). Some argue the need for measuring direct effects (Marshall 2010), while others suggest that habitat quality can be indirectly affected through the accumulation of multiple factors whose effects are not immediately visible. Ways of measuring this gradual impact are through increased energy costs (Riley 2008), higher instances of inbreeding (Melfi 2010), and even changes in the trophic level at which a population feeds (Gibson 2011).

Although anthropogenic habitat alteration typically results in lower quality food availability for nonhuman primates (Johns 1986; Felton et al. 2003), the behavioral and dietary adaptability of many macaque species allows them to persist in areas of disturbance (Riley 2007a). Such adaptability, however, may result in higher energetic costs that may not be adaptive in the long term. A group of Tonkean macaques in a heavily altered forest site was found to have daily path lengths similar to a second larger group inhabiting a relatively undamaged forest (Riley 2008). These relatively long daily paths for the small group were also confined to a smaller overall range, forcing the macaques to exploit the same parts of their home range more intensively than the other group (Riley 2008). Furthermore, home range adjustments become more difficult as high-quality habitats become smaller and spaced further apart (Melfi 2010). Small macaque populations are particularly vulnerable to habitat fragmentation because isolation from other groups decreases opportunities for gene flow and increases potential for the negative genetic effect of inbreeding (Melfi 2010). Disturbed or fragmented forest areas, however, should not be ignored by conservationists because incorporating mildly degraded forest areas into conservation strategies may be beneficial in terms of the future habitat carrying capacity, as has been suggested specifically for *Pongo pygmaeus morio* in Borneo (Marshall et al. 2006), but may be suitable for the highly adaptable macaque species' habitats as well (Paciulli 2010).

10.4.2 Crop Raiding

Anthropogenic habitat disturbance in the form of forest conversion for agricultural plots has a twofold effect on the habitat: it decreases the amount of forest habitat for the macaques and replaces it with a new and potentially appealing food source. As these forest-edge gardens can be located near to, or perhaps inside of, local macaque home ranges the monkeys can easily access the converted land to exploit the new resource in the area. Farmers see this as destructive behavior that is responsible for decreasing their crop yields and damaging their livelihoods. This forest-farm matrix scenario is common throughout Indonesia and represents a major source of tension. The forest-farm matrix can therefore be seen largely as both a *product* of anthropogenic habitat conversion and *source* of conflict with nonhuman primates through crop raiding.

The expansive range and large group size of macaques, along with their omnivorous, adaptable diet and high levels of intelligence are often cited as characteristics responsible for their success as crop raiders (Hill 2005; Lee and Priston 2005; Riley 2007b; Paciulli 2010). If macaque populations are frequent and successful crop raiders, this aspect of their behavior is likely to create the most powerful opinions and receive the strongest reactions from local humans. Often, farmers' tolerance of macaques is negatively correlated with high levels of *perceived* raiding: the more macaques are believed to destroy crops, the lower overall tolerance plantation owners feel for them. For example, in Lore Lindu National Park, Central Sulawesi, Riley (2007b) investigated the relationship between perceived levels of crop raiding and farmers' attitudes towards the local macaques. At this site cacao (*Theobroma cacao*) is consumed opportunistically by nearby Tonkean macaque populations (Riley 2007b). The large-bodied Tonkean macaques are conspicuous crop raiders that are not afraid of entering gardens, even when the farmer is present, and so have been characterized as detrimental to cacao production by many farmers in the area (Riley 2007b). The farmers in this study typically believed that macaques are the most frequent crop-raiding animal species and consume nearly 75% of their cacao crop (Riley 2007b). Quantitative measurements of crop losses, however, determined that the macaques were actually causing less crop damage than forest rats (Riley 2007b). This discrepancy may help foster more positive opinions regarding the Tonkean macaques, but it has to be understood by the local farmers if their opinions are going to change.

Many factors may influence farmers' perceptions of crop raiding, but the conspicuousness of potential raiders may be primary. Sumatran farmers who experience crop raiding from wild boars, pig-tailed macaques, and several other animal species incorrectly identified the wild boars as most destructive, when in fact it was the macaques (Linkie et al. 2007). Regardless of the actual damage to crops by macaques versus other animal species, humans are going to react according to their perceptions of the circumstances (Lee and Priston 2005; Riley 2007a). Therefore, in the future, it may be prudent to think about crop-raiding prevention strategies that are not only effective in lowering crop-raiding instances but are also effective at demonstrating actual levels of crop loss. Priston and Underdown (2009) have demonstrated how individual farmers can predict their risk of future crop loss by introducing a pest- and crop-specific formula that determines potential losses based on current losses. Risk is determined through an incidence rate that farmers can calculate by dividing the total number of damaged plants by the total number of plants at risk of being damaged (the sum of the damaged and undamaged plants) for a single crop species (Priston and Underdown 2009). This formula can be used throughout the year to establish seasonal variation in crop raiding (Priston and Underdown 2009), which is known to occur by crop type and pest species (Linkie et al. 2007). This simple method can assist farmers in choosing which crops to plant or not plant based on each crop's determined vulnerability to a particular pest (Priston and Underdown 2009). Other proposed strategies to mitigate crop raiding without harming macaques or other taxa include establishing buffer zones of preferred foods (Riley and Fuentes 2011) or physical barriers (Hockings and Humle 2009), taste aversion techniques, and guarding (Hill 2005).

10.4.3 Hunting

In Indonesia, macaques are hunted for a variety of reasons such as retaliation from crop raiding (Riley and Priston 2010), for consumption in ceremonial meals (Jones-Engel et al. 2005; Melfi 2010), to be kept as pets for a variety of different ends (Jones-Engel et al. 2005), and, more rarely, for medicinal application (Alves et al. 2010;

Peterson, unpublished data). Farmers have occasionally admitted to shooting raiding macaques opportunistically (Riley and Priston 2010), but in Sulawesi, they also frequently trap macaques to keep as pets (Jones-Engel et al. 2005). These pet macaques are typically either eaten or sold in the market for profit (Jones-Engel et al. 2005). Market-based hunting practices, as opposed to subsistence-based, are especially dangerous to protected animal species because the hunters are searching for higher economic gains through larger quantities of animals captured (Lee 1999). Research has suggested that demand for bushmeat in Sulawesi is on the rise and that hunting practices are becoming more sophisticated and intensified (Lee 1999; Lee et al. 2005; Melfi 2010). The critically endangered Sulawesi black-crested macaque is exceptionally vulnerable to increased hunting activity given its shrinking population and geographical proximity to the large bushmeat markets and human population densities (Melfi 2010). Bushmeat is also being brought to North Sulawesi from other provinces, causing animal species all over the island to be affected by this high demand (Lee et al. 2005). Large-scale markets like those in North Sulawesi may not be as prevalent across the rest of Indonesia due to many Muslim populations not participating in the capture and consumption of several wild animal species, such as monkeys (Lee et al. 2005). Hunting, however, does still occur in other parts of Indonesia such as Borneo (Marshall et al. 2006; Wadley and Colfer 2004), Java (Supriatna 2006), and Sumatra (Wheatley et al. 1999).

It has also been suggested that hunting pressures may actually be more damaging to nonhuman primate populations than other anthropogenic activities, such as habitat alteration (Marshall et al. 2006). Bornean orangutans (*P. pygmaeus morio*), whose slow life history patterns makes them inherently more susceptible to hunting, were found to have lower population densities when hunting pressures from nearby villages were present (Marshall et al. 2006). These population densities were more strongly correlated to hunting pressure proximity than logging intensity (Marshall et al. 2006). The hunting pressures experienced in this study were largely for private consumption or medicinal use (Marshall et al. 2006) and would fall under the category of "subsistence hunting," as opposed to the more intense commercial hunting that characterizes the bushmeat markets of North Sulawesi (Lee 1999). Therefore, unlike low- to mid-level anthropogenic habitat alteration, even low levels of hunting may be enough to negatively affect nonhuman primate population densities (Marshall et al. 2006). It is worth noting that low-intensity logging and forest fragmentation may have indirect effects on nonhuman primate populations by easing access into high-quality habitats for local hunters (Melfi 2010).

10.4.4 Disease Transmission

Researchers are increasingly interested in the epidemiological relationship between macaques and humans, especially in Indonesia. This interest may have been intensified by the recent discovery of the simian foamy virus (SFV), a zoonotic retrovirus, in both long-tailed macaques (*Macaca fascicularis*) and

Balinese temple employees (Engel et al. 2006). Although SFV is carried by Old World and New World monkeys, Balinese monkey temples have become primary locations for research on this pathogen as they experience a constant flow of tourists and employees who come into contact with monkeys on a daily basis (Engel et al. 2006). The risk for disease exchange between macaques and humans is significantly increased any time there are sustained interactions between the two species. Along with tourism activity, the pet trade has also been cited as a primary context of zoonotic disease transmission (Jones-Engel et al. 2005). The risk of catching diseases from macaques may result in widespread disregard for macaque populations in the future (Fuentes 2006).

It is important to note that close contact between human and nonhumans primates can have epidemiological implications for the macaque populations as well. Jones-Engel et al. (2004) found that in a large sample of pet macaques taken from across Sulawesi many were infected with intestinal parasites common in humans. These results suggest that the unique environment of pet monkeys, characterized by their high levels of contact with and reliance on nonprofessional human caretakers, causes their parasite load to differ substantially from that which is expected for wild macaque populations (Jones-Engel et al. 2004). Additionally, these anthropogenic parasite loads on pet macaques can be spread to wild populations through contact between the pets and wild individuals (Jones-Engel et al. 2005). Nonendemic pathogens can be introduced to wild macaque populations if the local pets are brought from a different island and contain nonendemic parasite loads (Jones-Engel et al. 2005). This transfer of novel pathogens may have important consequences for the health of wild macaque populations. Aside from disease transmission, there are physiological consequences for the macaques that experience frequent contact with humans such as increased stress levels, which bring with them consequences for individual macaque health as well as developing antisocial behavior towards humans (Lane et al. 2010).

Overall, there are many pathogens that can be cross transmitted between macaques and humans in Indonesia. The contexts in which zoonotic transmission are most frequently studied include Balinese monkey temples (Engel et al. 2006; Lane et al. 2011), markets (Malone et al. 2002), and among pet/performance monkeys (Jones-Engel et al. 2004; Schillaci et al. 2006). Below (Table 10.2), we have listed some of the most prominent pathogens that are shared among humans and macaques in Indonesia and the typical direction of transfer.

10.5 The Human-Macaque Interface: Cultural Facets

10.5.1 Mythology and Folklore

Where spatial overlap between human and nonhuman primates has a deep historical context, nonhuman primates may often be included in aspects of cultural mythology and folklore. An important aspect of the ethnoprimatological framework is focusing

Table 10.2 Bidirectional disease transmission

Pathogen	Type	Direction	Source
Entamoeba coli	Protozoon	Human → macaque	Jones-Engel et al. (2004)
Blastocystis hominis	Protozoon	Human → macaque	Jones-Engel et al. (2004)
Iodamoeba bütschlii	Protozoon	Human → macaque	Jones-Engel et al. (2004)
Entamoeba hartmanni	Protozoon	Human → macaque	Jones-Engel et al. (2004)
Endolimax nana	Protozoon	Human → macaque	Jones-Engel et al. (2004)
Chilomastrix mesnili	Protozoon	Human → macaque	Jones-Engel et al. (2004)
Ascaris spp.	Helminth	Human → macaque	Jones-Engel et al. (2004)
Trichuris spp.	Helminth	Human → macaque	Jones-Engel et al. (2004)
Hookworm	Helminth	Human → macaque	Jones-Engel et al. (2004)
Measles	Virus, respiratory	Human → macaque	Schillaci et al. (2006)
Rubella	Virus, respiratory	Human → macaque	Schillaci et al. (2006)
Parainfluenza 2 and 3	Virus, respiratory	Human → macaque	Schillaci et al. (2006)
Simian foamy virus	Retrovirus	Human ← macaque	Engel et al. (2006)
Herpesvirus B	Virus	Human ← macaque	Engel et al. (2002)

on situations where nonhuman primates have been included in the symbolic worldviews of humans and how that symbolism affects human attitudes towards them (Riley et al. 2011). Mythology often serves the dual purpose of entertainment and philosophical speculation, demonstrating the society's understanding of the ecological processes around them (Shepard 2002). When mythical tales focus on a specific animal species, anthropomorphizing the animal aids in demonstrating the storyteller's perspective regarding particular aspects of the human condition such as illness, death, and even desirable or undesirable personality traits (Shepard 2002). Nonhuman primates are ideal characters in mythology due to their readily observable behaviors and habits which can easily be interpreted as analogues to aspects of human behavior (Mullin 1999).

One outcome of folklore and mythology can be the local conservation of a particular species due to preexisting taboos against harming them. Taboos are often associated with folklore and mythology and protect species by virtue of cultural custom (Colding and Folke 2001; Saj et al. 2006). In contrast to formally prescribed prohibitions, informal self-imposed and self-monitored cultural conventions or norms protect specific habitats by regulating adherents' interactions with the environment in terms of resource access and use (Colding and Folke 2001). Habitat taboos are often associated with religious or spiritual perceptions of sacredness that facilitate the habitat's protection (Colding and Folke 2001). Plant and animal species within the habitat are often protected under these taboos by extension, through bans on hunting, fishing, and harvesting any of the resources therein (Colding and Folke 2001).

Forest patches receiving protection under such taboos are commonly referred to as *sacred groves* (Baker et al. 2009; Colding and Folke 2001). Often located within these sacred groves are shrines possessing a religious or spiritual significance that is then applied to the entire area surrounding them (Baker et al. 2009; Colding and Folke 2001). Although these sacred forest sites are of spiritual and ritual importance that can occasionally contribute to habitat conservation (Baker et al. 2009; Saj et al. 2006), rich wildlife diversity within them may also present hunting opportunities for local residents (Wadley and Colfer 2004). Aside from sacred grove hunting, perceptions of sacredness themselves are subject to change over time as cultural groups can outgrow the taboo which a conservation program has attempted to bind them to (Baker et al. 2009). Furthermore, recent migrants to an area will not typically share local peoples' cultural taboos and therefore freely hunt, trap, or kill a "protected" species if the opportunity arises (Colquhoun 2005; Riley 2007b). Riley's (2007b) research on macaque folklore in Central Sulawesi, Indonesia suggests that differing cultural beliefs between separate ethnic groups complicates the idea of structuring permanent conservation programs on the customs of one group when others have equal access to the same resources. For example, recent migrants to Lore Lindu National Park do not share the same level of reverence for local Tonkean macaques as some of the indigenous members of the community (Riley 2010). This difference is attributed largely to the migrants' lack of cultural associations with Tonkean macaques. Additionally, temporary migrants to an area may have a deleterious effect on local ecology because they are less inclined to pursue sustainable patterns of resource use during their short-term occupation (Hill 2005).

Mythology and folkloric knowledge can also instruct people as to which nonhuman primate species they *should* consume. Many species in Indonesia are reportedly used in traditional folk medicine (Alves et al. 2010). For instance, in parts of Indonesia, macaque liver is believed to cure asthma, while the flesh, though not having medicinal properties, is enjoyed for its uniquely "hot" flavor (Peterson, unpublished data). The existence of these cultural preferences for nonhuman primate consumption adds complexity to the human-macaque interface in Indonesia, where the diversity of cultural practices, mythology, and folklore can result in species protection as well as exploitation.

10.5.2 Religion

Much of the work regarding the religious component of the human-macaque interface in Indonesia has taken place in the Balinese monkey forests and sacred temple sites. The current interspecies relationships displayed in these locations are representative of the extensive history of sympatry between humans and long-tailed macaques (*Macaca fascicularis*) on Bali (Loudon et al. 2006; Wheatley 1999). The sacred forests around the *Pura Dalem*, or funerary temples, are colloquially known as "monkey forests" when the macaques inhabit them (Fuentes et al. 2005; Wheatley 1999). Like the forests, monkeys associated with these temple sites may be considered sacred and offered a degree of protection as a result (Fuentes et al. 2005; Loudon et al. 2006; Wheatley 1999). Many of the larger monkey forest temples provision the local macaques, keeping them in the area to serve the additional purpose of tourist attraction (Fuentes et al. 2007). Fuentes et al. (2007) believe that along with religious taboos, tourism may be advantageous to the macaques by providing them with additional food sources from the tourists and a decreased risk of predation. These advantages for the macaques are mirrored by the increased economic incentive for the local inhabitants to keep them protected as well (Fuentes et al. 2007). A potentially negative result of increased tourism is disease transmission between macaques and humans (Engel et al. 2006; Fuentes 2006). Diseases that move from macaques to humans may end up fostering negative opinions of the sacred macaques in these tourist sites.

The integration of temple sacredness with the surrounding environment on Bali is embedded in the unique form of Hinduism practiced there. Balinese monkey forests are characterized by the philosophical concept of *Tri Hita Karana*, which states that the production of goodness and well-being is only possible through harmonious interaction between the three elements of the world: God, man, and nature (Jensen and Suyrani 1992; Wheatley 1999). The Balinese funerary temples and the associated monkey forests represent a convergence point for the aforementioned religious, ecological, and economic factors resulting in taboos on harming the macaques and their habitat. It is here, within these temple sites and sacred grounds, that interactions between the Balinese people and macaques have largely been examined (e.g., Wheatley 1999; Loudon et al. 2006; Fuentes et al. 2007).

10.6 The Paradox of Macaque Sacredness

Recently, the characterization of "ubiquitous sacredness" among Balinese macaques and their protected status has been called into question (Loudon et al. 2006; Schillaci et al. 2010). These authors suggest that sacredness may only be applied to long-tailed macaques in sacred temple spaces and that when found outside of this context, they are treated as pest animals. This may be related to their economic significance in Balinese temples where tourism has become a profitable enterprise in contrast to potential profit loss through crop raiding by these same macaques in a different spatial context (Fuentes et al. 2007; Schillaci et al. 2010). Drawing upon this important notion of spatial context are inquiries into Balinese perceptions of macaque sacredness outside of Bali as well. Previous research has suggested that Balinese transmigrants in Sulawesi do not regard the local macaques as sacred (Jones-Engel et al. 2005). These conclusions, however, have not yet been systematically investigated in a study of their own.

The Hindu population of Sulawesi is derived almost exclusively from Balinese transmigrants (Davis 1976; Whitten et al. 1987). Many Balinese people moved to Sulawesi as part of the transmigration program instituted by the Indonesian government in the 1950s which officially extended into the 1980s (Davis 1976; World Bank 1988). As a result, thousands of Balinese families have relocated to other areas in Indonesia, including the eastern Indonesian island of Sulawesi (Davis 1976; Whitten et al. 1987). The Balinese transmigrants in Sulawesi now reside in transmigrant communities where their religious and ethnic identities have remained intact (Jones-Engel et al. 2005). The transmigrant areas, however, vary geographically and ecologically from the small mountainous island of Bali, which may influence transmigrant perceptions of their new space. The new environmental surroundings also lack Bali's long history of human occupation, especially regarding the incorporation of religiously symbolic landscapes and ancient temple sites. All Balinese Hindu temples in Sulawesi have been constructed by the transmigrants upon their arrival. Therefore, any reverence that is observed for Sulawesi macaques by the Balinese transmigrants need not be the result of their long-standing occupation of sacred grounds. Instead, this could be a feeling that was brought with them from Bali and applied to the new transmigrant space in Sulawesi.

One of us (Peterson) is currently exploring whether the relocation of Balinese transmigrants to South Sulawesi has affected their perceptions of macaque sacredness and, if so, to what degree does that alter the two species' interactions and coexistence. The major ethnic groups known to inhabit the research area (Luwu Timor district) are the Toraja Kaili, Pamona, and the coastal Bugis – Makassar Bugis (Davis 1976; Whitten et al. 1987). Through transmigration, families from Bali were relocated to the Kalaena transmigrant settlement (Roth 2009). Transmigrants from Java are also located within this region, and villages are comprised of solely Balinese or Javanese transmigrants, or are a mixed population with native residents as well (Roth 2009).

Preliminary results support the suggestion that perceptions of macaque sacredness are, indeed, more strongly tied to space than to an inherent holiness of monkeys in the eyes of Balinese Hinduism (Jones-Engel et al. 2005; Schillaci et al. 2010). After responding that the local booted macaques (*Macaca ochreata*) were not considered sacred, many of the transmigrants specified that this is because they do not live in temple sites like the macaques in Bali (Peterson, unpublished data). Therefore, the respondents in this study appear to make important distinctions between forest monkeys and temple monkeys when asked to explore their own perceptions of macaque sacredness.

One issue closely tied to this spatial factor is the level of macaque habituation. The fact that these booted macaque populations live in the forest-farm matrix neighboring transmigrant villages causes them to come into far less contact with humans than the long-tailed macaques that reside in the urban/temple environment of Bali. When discussing interactions with *M. ochreata*, respondents often lamented that these monkeys are too afraid of humans and always run away if they meet (Peterson, unpublished data). This is not surprising as the majority of these interactions occur when villagers enter the forest to search for wood, or with dogs to hunt pigs or monkeys. These interactional contexts are not conducive to habituation, and the lack of a centralized living space for the macaques like the temple complexes in Bali makes it difficult to conceive of the booted macaques becoming habituated to local villagers.

Along with the fearful behaviors of unhabituated macaques, crop raiding has been cited as one of the primary behavioral characteristics preventing the booted macaques from being considered sacred (Peterson, unpublished data). These two behavioral issues (crop raiding and unhabituated behaviors) are inevitably linked together as crop raiding can cause response hunting, which in turn reinforces the unhabituated behaviors. Included in our definition of "unhabituated behaviors" are things listed by the respondents such as running away from humans, not "wanting" or being able to adapt to humans, and being untame or "wild" in general (Peterson, unpublished data). The relevance of macaque behavior for its effect on perceptions of macaque sacredness is supported by recent research suggesting that even the long-tailed macaques in Bali are dissuaded from crop raiding by farmers with the use of pellet guns (Schillaci et al. 2010). Additionally, *M. fascicularis* are often chased out of gardens and even hunted and eaten near temple areas less frequented by tourists where the macaques are less habituated to humans and unprotected (Loudon et al. 2006). These findings suggest that even in Bali there is no ubiquitous sacredness for temple monkeys if they begin to interfere with farmers' livelihoods in the forest-farm matrix. This same principle seems to hold true for the Balinese transmigrants and their relationship with the booted macaques of Sulawesi.

10.7 Conclusions

In this chapter, we describe the multifaceted nature of the human-macaque interface in Indonesia. We see this interface as being divided into three major environmental settings: urban/temple, protected forest area, and forest-farm matrix. As we have established, these environmental settings often overlap or appear contiguously, and therefore, a number of the Indonesian macaque species are associated with more than one of these environmental contexts. This ecological diversity presents a dynamic spatial setting for the human-macaque interface, with intricacies that should be addressed in future ethnoprimatological inquiries. Riley (2006) suggests incorporating primatological methodology to record both aspects of macaque behavioral ecology and the extent of their interaction with local humans, as well as ethnographic methodology to understand why humans believe these interactions are taking place and the cultural importance attributed to them. This mixed methodologies approach can help uncover the nuanced relationships between humans and macaques, specifically how these relationships vary throughout the different environmental contexts in which they occur.

Primatological studies that focus on the human-macaque interface are as timely now as ever and will continue to be relevant in the future. The increasing human population in Indonesia and around the world is resulting in expanded contexts for contact between human and nonhuman primates, providing indefinite opportunities for new inquiries. Knowledge gained through investigations into the human-macaque interface has both theoretical and applied significance (Riley and Fuentes 2011). Theoretically, such explorations contribute to our understanding of how human behavior shapes the socio-ecological pressures that act upon primates and other organisms across landscapes and ecosystems (Riley and Fuentes 2011). One example is the increasingly studied epidemiological landscape of the human-macaque interface. It has been demonstrated that increased contact between macaques and humans can introduce novel zoonotic pressures on wild macaque populations through direct contact with humans (Engel et al. 2006; Fuentes 2006) or indirectly through pets (Jones-Engel et al. 2005). Studying the intricacies of this epidemiological landscape can address the situations in which both macaques and humans are most at risk of disease transmission and how these risks can best be mitigated.

Another future direction might be to explore how the predominant context of the human-macaque interface, the forest-farm matrix, and specifically access to cultivated foods is shaping macaque biology and behavior. Significant dietary changes from forest to cultivated foods that would otherwise be inaccessible in pristine forest (i.e., cacao, corn, cassava) may affect various aspects of macaque biology, health, behavior, and ultimately, fitness. Some of the behavioral changes that may accompany this dietary shift include traveling and foraging patterns. The range over which macaques traverse may be altered due to the inclusion of cultivated lands. Also, larger macaque groups may split off into smaller "foraging parties" to avoid detection from local humans when they enter the cultivated foraging grounds. Foraging for primarily cultivated foods may require other adaptations to avoid human aggression, such as foraging at times when fields are free of people or carrying food off-site to eat in safety later. Macaque cheek pouches may be especially helpful in facilitating the latter of these potential behavioral adaptations. Incorporating more cultivated food sources into macaque diets may also provide opportunities for changes to occur in their communication systems that are relevant to the new feeding context.

The applied contributions of inquiries into the human-macaque interface are largely in the area of conservation. The long-term holistic studies we suggest in this chapter can integrate conservation approaches that are known to be appropriate for a given environmental setting and tailor them to the specific cultural and ecological facets of the new area. Using informal institutions (e.g., habitat taboos) for the basis of conservation programs may be effective in this light, but it comes with caveats. One caveat, as outlined by Riley (2010), is the potentially narrow scope of taboos that may only apply to one specific animal species or a small patch of land. These taboos may not be easily extended to encompass the entire ecosystem in which the taboo is upheld (Riley 2010). Furthermore, as taboos are typically specific to a single ethnic group, the migration of a different one into the area may make the established conservation program based on local taboos difficult to enforce (Colquhoun 2005; Riley 2010). Therefore, future conservation-oriented research should describe the effective and ineffective aspects of informally protected resource and habitat taboos to add depth to our understanding of the cultural factors affecting habitat protection. As previously demonstrated, the spatial context in which macaques and the Balinese interact appears to be fundamental in guiding the cultural and ecological manifestations of their relationship. In the forest-farm matrix, cultural notions of macaque sacredness are sometimes forgotten in the interest of protecting farmers' crops, which indicates an important distinction between protected spaces for symbolically religious reasons and spaces that need to be protected for subsistence purposes. Because the potential for conflict is much higher in the forest-farm matrix than the other environmental contexts, applied conservation programs will benefit from understanding the difficulties and potential conservation roadblocks that come as a result of this forest-farm matrix conflict.

The intersection of the ecological and cultural facets that comprise the human-macaque interface can be thought of as a matrix whose constituent components fluctuate in size, frequency, and influence from place to place. The capacity for culture to change through time, adjusting to changes or perceived changes in the ecological or cultural surroundings, presents a complex situation to be understood by the primatologist. The lines between what constitutes a cultural or ecological facet also become blurred when we think about the economic goals that are behind logging and the quest for livelihood that inspires some forms of forest conversion for new agricultural land. Because these facets are so fluid and interrelated, long-term applications of the mixed methodologies approach will likely be the best way to document how relationships within the human-macaque interface shift, resulting in conflict where there was once commensalism, mutualism where there was once conflict, and the whole gamut of potential relationship manifestations.

Acknowledgments Thanks to the Indonesian Ministry of Research and Technology (RISTEK) for granting permission to conduct the research presented in this chapter and to Prof. Putu Oka for sponsoring this research. We are extremely grateful to the Fulbright US Scholar Program for providing funding. We also thank an anonymous reviewer for their helpful comments on earlier drafts. Special thanks to Pak Redo, Pak Donald, Pak Nyoman Santosa, and Pak Charles for their hospitality and support while conducting this research.

References

Abegg C, Thierry B (2002) Macaque evolution and dispersal in insular South-East Asia. Biol J Linn Soc 75:555–576

Alves RRN, Souto WSS, Barboza RRD (2010) Primates in traditional folk medicine: a world overview. Mamm Rev 40:155–180

Atkinson JM (1983) Religions in Dialogue: The Construction of an Indonesian Minority Religion. American Ethnologist 10(4):684–696.

Baker LR, Tanimola AA, Olubode OS, Garshelis DL (2009) Distribution and abundance of sacred monkeys in Igboland, southern Nigeria. Am J Primatol 71:574–586

Bishop N, Hrdy SB, Teas J, Moore J (1981) Measures of human influence in habitats of south Asian Monkeys. Int J Primatol 2:153–167

BPS (2000) Biro Pusat Statistik 2000 population census. BPS, Bogor

CIA (2011) The world factbook: Indonesia. Downloaded from https://www.cia.gov/library/publications/the-world-factbook/geos/id.html

Colding J, Folke C (2001) Social taboos: "Invisible" systems of local resource management and biological conservation. Ecol Appl 11:584–600

Colquhoun IC (2005) Primates in the forest: Sakalava ethnoprimatology and synecological relations with black lemurs at Ambato Massif, Madagascar. In: Paterson JD, Wallis J (eds) Commensalism and conflict: the human-primate interface. American Society of Primatologists, Norman

Davis GJ (1976) Parigi: a social history of the Balinese movement to Sulawesi, 1907–1974. PhD thesis, Stanford University, Palo Alto

Engel G, Hungerford LL, Jones-Engel L, Travis D, Eberle R, Fuentes A, Grant R (2006) Risk assessment: a model for predicting cross-species transmission of Simian Foamy Virus from macaques (*M. fascicularis*) to humans at a monkey temple in Bali, Indonesia. Am J Primatol 68:934–948

Engel GA, Jones-Engel L, Schillaci MA, Suaryana KG, Putra A, Fuentes A, and Henkel R (2002) Human Exposure to Herpesvirus B-Seropositive Macaques, Bali, Indonesia. Emerging Infectious Diseases 8(8):789–795

Fuentes A (2006) Human culture and monkey behavior: assessing the contexts of potential pathogen transmission between macaques and humans. Am J Primatol 68:880–896

Felton AM, Engstrom LM, Felton A, and Knott CD (2003) Orangutan population density, forest structure and fruit availability in hand-logged and unlogged peat swamp forests in West Kalimantan, Indonesia. Biological Conservation 114:91–101

Fuentes A, Southern M, Suaryana KG (2005) Monkey forests and human landscapes: is extensive sympatry sustainable for *Homo sapiens* and *Macaca fascicularis* on Bali? In: Paterson JD, Wallis J (eds) Commensalism and conflict: the human-primate interface. American Society of Primatologists, Norman

Fuentes A, Shaw E, Cortes J (2007) Qualitative assessment of macaque tourist sites in Padangtegal, Bali, Indonesia, and the Upper Rock Nature Reserve, Gibraltar. Int J Primatol 28:1143–1158

Gibson L (2011) Possible shift in macaque trophic level following a century of biodiversity loss in Singapore. Primates 52:217–220

Hill CM (2005) People, crops, and primates: a conflict of interest. In: Paterson JD, Wallis J (eds) Commensalism and conflict: the human-primate interface. American Society of Primatologists, Norman

Hockings K, Humle T (2009) Best practice guidelines for the prevention and mitigation of conflict between humans and great apes. Occas Pap Species Surviv Comm 37:41

Hoey BA (2003) Nationalism in Indonesia: building imagined and intentional communities through transmigration. Ethnology 42:109–126

IUCN (2010) IUCN red list of threatened species. www.iucnredlist.org

Jensen GD, Suyrani LK (1992) The Balinese people: a reinvestigation of character. Oxford University Press, New York

Johns AD (1986) Effects of selective logging on the behavioral ecology of west Malaysian primates. Ecology 67:684–694

Jones-Engel L, Engel GA, Schillaci MA, Kyes K, Froehlich J, Paputungan U, Kyes RC (2004) Prevalence of enteric parasites in pet macaques in Sulawesi, Indonesia. Am J Primatol 62:71–82

Jones-Engel L, Shillaci MA, Engel G, Paputungan U, Froehlich JW (2005) Characterizing primate pet ownership in Sulawesi: implications for disease transmission. In: Paterson JD, Wallis J (eds) Commensalism and conflict: the human-primate interface. American Society of Primatologists, Norman

Lane KE, Lute M, Rompis A, Wandia IN, Arta Putra IGA, Hollocher H, Fuentes A (2010) Pests, pestilence, and people: the long-tailed macaque and its role in the cultural complexities of Bali. In: Gursky-Doyen S, Supriatna J (eds) Indonesian primates. Springer, New York, pp 235–248

Lane KE, Holley C, Hollocher H, Fuentes A (2011) The anthropogenic environment lessens the intensity and prevalence of gastrointestinal parasites in Balinese long-tailed macaques (*Macaca fascicularis*). Primates 52:117–128

Lee RJ (1999) Market hunting pressures in north Sulawesi, Indonesia. Trop Biodivers 6:145–162

Lee PC, Priston NEC (2005) Human attitudes to primates: perceptions of pests, conflict and consequences for primate conservation. In: Paterson JD, Wallis J (eds) Commensalism and conflict: the primate-human interface. American Society of Primatologists, Norman, pp 1–23

Lee RJ, Gorog AJ, Dwiyahreni A, Siwu S, Riley J, Alexander H, Paoli GD, Ramono W (2005) Wildlife trade and implications for law enforcement in Indonesia: a case study from north Sulawesi. Biol Conserv 123:477–488

Linkie M, Dinata Y, Nofrianto A, Leader-Williams N (2007) Patterns and perceptions of wildlife crop raiding in and around Kerinci Seblat National Park, Sumatra. Anim Conserv 10:127–135

Loudon JE, Howells ME, Fuentes A (2006) The importance of integrative anthropology: a preliminary investigation employing primatological and cultural anthropological data collection methods in assessing human-monkey co-existence in Bali, Indonesia. Ecol Environ Anthropol 2:2–13

Malone N, Purnama AR, Wedana M, Fuentes A (2002) Assessment of the sale of primates at Indonesian bird markets. Asian Primates 8:7–11

Marshall AJ (2010) Effect of habitat quality on primate populations in Kalimantan: Gibbons and leaf monkeys as case studies. In: Gursky-Doyen S, Supriatna J (eds) Indonesian primates. Springer, New York, pp 157–177

Marshall AJ, Nardiyono, Engstrom LM, Pamungkas B, Meijaard JP, Stanley SA (2006) The blowgun is mightier than the chainsaw in determining population density of Bornean orangutans (*Pongo pygmaeus morio*) in the forests of east Kalimantan. Biol Conserv 129:566–578

Melfi V (2010) Selamatkan yaki! Conservation of Sulawesi crested black macaques *Macaca nigra*. In: Gursky-Doyen S, Supriatna J (eds) Indonesian primates. Springer, New York, pp 343–357

Mullin MH (1999) Mirrors and windows: sociocultural studies of human-animal relationships. Annu Rev Anthropol 28:201–224

Paciulli LM (2010) The relationship between nonhuman primate densities and vegetation on the Pagai, Mentawai Islands, Indonesia. In: Gursky-Doyen S, Supriatna J (eds) Indonesian primates. Springer, New York, pp 199–215

Priston NEC, Underdown SJ (2009) A simple method for calculating the likelihood of crop damage by primates: an epidemiological approach. Int J Pest Manag 55:51–56

Riley EP (2006) Ethnoprimatology: toward reconciliation of biological and cultural anthropology. Ecol Environ Anthropol 2:75–86

Riley EP (2007a) Flexibility in diet and activity patterns of Macaca tonkeana in response to anthropogenic habitat alteration. Int J Primatol 28:107–133

Riley EP (2007b) The human-macaque interface: conservation implications of current and future overlap and conflict in Lore Lindu National Park, Sulawesi, Indonesia. Am Anthropol 109:473–484

Riley EP (2008) Ranging patterns and habitat use of Sulawesi Tonkean macaques (*Macaca tonkeana*) in a human-modified habitat. Am J Primatol 70:670–679

Riley EP (2010) The importance of human–macaque folklore for conservation in Lore Lindu National Park, Sulawesi, Indonesia. Oryx 44:235–240

Riley EP, Priston NEC (2010) Macaques in farms and folklore: exploring the human–nonhuman primate interface in Sulawesi, Indonesia. Am J Primatol 71:1–7

Riley EP, Fuentes A (2011) Conserving social-ecological systems in Indonesia: human-nonhuman primate interconnections in Bali and Sulawesi. Am J Primatol 73:62–74

Riley EP, Fuentes A, Wolfe LD (2011) Ethnoprimatology: contextualizing human and nonhuman primate interactions. In: Campbell C, Fuentes A, MacKinnon K, Bearder S, Stumpf R (eds) Primates in perspective. Oxford University Press, New York, pp 676–686

Roth D (2009) Property and authority in a migrant society: Balinese irrigators in Sulawesi, Indonesia. Dev Change 40:195–217

Saj TL, Mather C, Sicotte P (2006) Traditional taboos in biological conservation: the case of *Colobus vellerosus* at the Boabeng-Fiema Monkey Sanctuary, central Ghana. Soc Sci Inf 45:285–310

Schillaci MA, Jones-Engel L, Engel G, Kyes R (2006) Exposure to human respiratory viruses among urban performing monkeys in Indonesia. Am J Trop Med Hyg 75:716–719

Schillaci MA, Engel GA, Fuentes A, Rompis A, Arta Putra IGA, Wandia IN, Bailey JA, Brogdon BG, Jones-Engel L (2010) The not-so-sacred monkeys of Bali: a radiographic study of human-primate commensalism. In: Gursky-Doyen S, Supriatna J (eds) Indonesian primates. Springer, New York

Schiller A (1996) An "old" religion in "new order" Indonesia: notes on ethnicity and religious affiliation. Soc Relig 57:409–417

Shepard GH (2002) Primates in Matsigenka: subsistence and world view. In: Fuentes A, Wolfe L (eds) Primates face to face: the conservation implications of human-nonhuman primate interconnections. Cambridge University Press, Cambridge, pp 101–136

Supriatna J (2006) Conservation programs for the endangered Javan gibbon (*Hylobates moloch*). Primate Conserv 21:155–162

Wadley RL, Colfer CJP (2004) Sacred forest, hunting, and conservation in west Kalimantan, Indonesia. Hum Ecol 32:313–338

Wheatley B (1999) The sacred monkeys of Bali. Waveland Press, Prospect Heights
Wheatley BP, Stephenson R, Kurashina H (1999) The effects of hunting on the longtailed macaques of Ngeaur Island, Palau. In: Dolhinow P, Fuentes A (eds) The nonhuman primates. Mayfield, Mountain View, pp 159–163
Whitten AJ, Mustafa M, Henderson GS (1987) The ecology of Sulawesi. Gadjah Mada University Press, Yogyakarta
World Bank (1988) Indonesia: The Transmigration Program in Perspective. A World Bank Country Study. Washington: The World Bank
WRI (2006) Biodiversity and protected areas – country profile: Indonesia. Downloaded from http://earthtrends.wri.org/text/biodiversity-protected/country-profile-86.html

Chapter 11
Out of Asia: The Singular Case of the Barbary Macaque

Bonaventura Majolo, Els van Lavieren, Laëtitia Maréchal, Ann MacLarnon, Garry Marvin, Mohamed Qarro, and Stuart Semple

11.1 Introduction

The case of the Barbary macaque (*Macaca sylvanus*) is a rather interesting one for several reasons. For one, it is the only species of the genus *Macaca* that is found outside Asia. Its ancient lineage and present fragmented distribution in Africa, so strikingly afar from the other macaque species, place it central to all discussions on macaque phylogeography. It is also one of a handful of primate species that live in a cold/snowy environment for a significant part of the year. For these reasons, the species represents an excellent model to study macaque evolution and dispersal and primate adaptation to temperate/cold climates. Currently, two disjointed and highly fragmented populations of this species exist in the wild, in Morocco and Algeria. Outside Africa, a free-ranging population of macaques inhabits the Rock of Gibraltar. The species is probably the most intensively studied macaque after rhesus (*M. mulatta*) and Japanese (*M. fuscata*) macaques. However, much of this body of research is based on the small and highly provisioned Gibraltar population or on free-ranging captive populations in three parks in Germany and France. Fewer studies have been conducted on wild/non-provisioned Barbary macaques.

B. Majolo (✉)
School of Psychology, University of Lincoln, Brayford Pool, Lincoln, LN6 7TS, UK
e-mail: bmajolo@lincoln.ac.uk

E. van Lavieren
Moroccan Primate Conservation Foundation, Erfstraat 23, 6668 AD Randwijk, Amsterdam, the Netherlands

L. Maréchal • A. MacLarnon • G. Marvin • S. Semple
Centre for Research in Evolutionary and Environmental Anthropology, University of Roehampton, Holybourne Avenue, SW15 4JD London, UK

M. Qarro
Ecole Nationale Forestiére d'Ingénieurs, Salé, Hassane Rabat, Morocco

S. Radhakrishna et al. (eds.), *The Macaque Connection: Cooperation and Conflict between Humans and Macaques*, Developments in Primatology: Progress and Prospects 43,
DOI 10.1007/978-1-4614-3967-7_11,

The Barbary macaque is listed as endangered in the IUCN 2010 Red List of Threatened Species, and the total population size in the wild is estimated at between 5,000 and 6,000 individuals. The rapid decline in population numbers in the recent decades and the increasingly negative impacts of human interactions on the species make it vital that conservation measures are strengthened at the earliest opportunity. In this chapter, we first review the literature on wild Barbary macaques to describe their geographic distribution, ecology and behaviour. We then briefly describe the nature and frequency of interactions between Barbary macaques and humans in both historic and present times and discuss factors threatening the survival of this species. Finally, as a case study of human-macaque interactions, we analyse the effects of tourist pressure on the behaviour of Barbary macaques at our field site in the Middle Atlas Mountains of Morocco.

11.2 The Geographic Distribution and Socioecology of the Barbary Macaque: A Short Overview

In historic times, the Barbary macaque was an inhabitant of parts of Europe and all of North Africa, from Libya and Egypt to Morocco (Delson 1980; Camperio Ciani 1986). Its current distribution is limited to small relict patches of forest and scrub, in the rocky and mountainous parts of the Rif and Atlas Mountains (Morocco) and in parts of the Tellian Atlas in Algeria (Fa 1984a; Camperio Ciani 1986; Ménard and Vallet 1993; Scheffrahn et al. 1993). As such, the Barbary macaque is the northernmost African non-human primate. The current distribution of this species ranges from a latitude of around 31° 15′N to 36° 45′N and from a longitude of around 7° 45′W to 5° 35′E (Fooden 2007). A small free-ranging and provisioned population of around 200 macaques inhabits the Upper Rock Nature Reserve of Gibraltar. Being the only African macaque species, the distribution of the Barbary macaque is far away from the other extant macaque species in southeast Asia, from Pakistan and India in the west to Japan in the north and Indonesia in the southeast. The geographic distribution of the various *Macaca* species reflects their phyletic origin (Abegg and Thierry 2002). The Barbary macaque belongs to the *silenus-sylvanus* group, from which three other phyletic groups *fascicularis*, *sinica*, and *arctoides* evolved (Fooden 1976; Abegg and Thierry 2002).

As mentioned above, the Barbary macaque is now an endangered species. The species density estimates of the last few decades suggest a dramatic and steady decrease in the number of macaques in Algeria and Morocco. The total population size decreased from an estimated 21,000 individuals around 40 years ago (Taub 1975) to 10,000–16,000 in the early 1990s (Lilly and Mehlman 1993; von Segesser et al. 1999; Table 11.1). At present, the total population is estimated to be between 5,000 and 6,000 (van Lavieren and Wich 2009; Siân Waters, personal communication). More detailed and frequent population surveys have been conducted in Morocco, particularly in the Middle Atlas Mountains (e.g. Deag 1977; Camperio

Table 11.1 Population and density estimates for the Barbary macaque since 1968

Location	Estimate *N* monkeys/km²	Year of survey	Reference
MAM,[a] Morocco	70	1968	Deag (1974, 1984)
	43	1977	Taub (1977)
	28	1994	Camperio Ciani et al. (2005)
	≥7	2002	Camperio Ciani et al. (2005)
	≥12.1	2005	van Lavieren and Wich (2009)
	Population size (*N* monkeys)		
MAM, Morocco	4,000–5,000	2005	van Lavieren and Wich (2009)
Morocco	17,000	1974	Taub (1975)
	6,000–10,000	Unknown	Ross (2004)
Total population	≤21,500	1974	Taub (1975)
	15,000	Unknown	von Segesser et al. (1999)
	5,000–6,000	2009	van Lavieren, unpublished

[a]*MAM* Middle Atlas Mountains

Ciani et al. 2005; van Lavieren and Wich 2009), but the decline is also evident in the Algerian populations (von Segesser et al. 1999). In the early 1970s, Deag (1984) estimated a macaque density of around 70 individuals per km² in the Middle Atlas Mountains. By 2002, the density had already dropped to between seven and ten individuals per km², and in 2005, the average number of monkeys in the Ifrane National Park was as low as seven individuals per km² (Camperio Ciani et al. 2005; Table 11.1). This is a very low figure and is due to habitat fragmentation, competition over food sources between macaques and humans and their livestock and the illegal capturing of infants for the pet trade. With such a small population size, the survival of the species is even more uncertain, considering the fact that both the Algerian and Moroccan sub-populations are small and highly fragmented, with large gaps (up to around 700 km) existing between them. For example, in the sub-population living in the Ifrane National Park, in the Middle Atlas Mountains, fragmentation of the forested areas prevents migration between groups and poses a threat to their survival. Moreover, a recent study showed that wild populations are genetically distinct from one another and show low overall genetic diversity (Modolo et al. 2005). This poses a major threat for population viability, as small, fragmented sub-populations with low genetic diversity are clearly more vulnerable to factors such as human disturbance and infectious diseases. For example, the introduction of a new disease to the small Gibraltar population of Barbary macaques could quickly eliminate the whole population (Fa and Lind 1996; Honess et al. 2006). Such a dramatic event could also occur in the wild populations, as interactions between macaques and humans are frequent in both Morocco and Algeria. This highlights the need for effective conservation plans to be designed and implemented urgently, to save those populations.

The Barbary macaque is one of very few primates that live in temperate cold climates (Hanya et al. 2011). In its native habitat, the species is found in cedar-oak (*Cedrus atlantica* and *Quercus ilex*) and deciduous oak forests (*Q. faginea* and *Q. afares*), scrub, grassland and rocky ridges (Fooden 2007). It can be considered a flagship species of the cedar and oak forests of Morocco and Algeria, an ecosystem with a high degree of biodiversity and a number of rare and endemic plant and animal species. The elevation distribution of the Barbary macaque is considered to lie at altitudes between 400 and 2,300 m (Fooden 2007). As such, this species experiences strong seasonal differences, from dry and hot summer months, when temperatures can reach up to 40 °C, to wet and snowy winter months with temperatures often going below 0 °C (Majolo et al. under review). As a possible consequence of such strong seasonality and the limited time availability of some food sources, the Barbary macaque is a very eclectic forager. Its diet comprises leaves, seeds, fruit, fungi, plant and animal matter, sap and roots (Ménard 2002).

Sexual dimorphism can be observed in relation to body length (males: 550–600 mm, females: 450 mm) and weight (males: 15–17 kg, females: 10–11 kg; Fa 1989). Furthermore, males have elongated canines that play an important role in dominance-related behaviours. The maximum recorded lifespan is 28 years in captivity (Sue Wiper, personal communication 2011; Fooden 2007), but monkeys are unlikely to live beyond 15–17 years in the wild. Sexual maturity is reached at around 6 years of age in males and 5 years of age in females (Ménard et al. 1985). The Barbary macaque is a seasonal breeder with a mating season occurring between October and January and births occurring between late March and July/August (Dixon 1998). Similar to other macaque species, in the mating season females show significant perineal swellings, during which ovulation occurs (Möhle et al. 2005). In the wild, Barbary macaques live in multimale-multifemale groups of 10–60 individuals with a sex ratio ranging between 0.6 and 1.6 females per male (Ménard 2002). Conversely, the provisioned groups of Gibraltar are composed of around 60–80 individuals. In Morocco and Algeria, group size and composition can vary dramatically, probably due to differences across seasons and years in food abundance. In Morocco, the average group size in the Ifrane National Park is around 15–25 individuals, whereas it might be higher in the Rif Mountains (Siân Waters, personal communication 2011). The Barbary macaque is considered a relatively tolerant species (i.e. Grade III) according to the classification proposed by Thierry (2000). This means that group members should form a shallow dominance hierarchy and conflicts should be often undecided and/or result in counter-aggression to the aggressor by the recipient (Thierry 2007). Moreover, Barbary macaques should have a low degree of kin bias in grooming distribution and a high frequency of reconciled conflict at the group level (Thierry 2000). However, this picture may not fully represent wild populations. For example, two recent studies (McFarland and Majolo 2011a, b) found that 5% of conflicts were undecided and less than 4% resulted in counter-aggression in two wild groups of macaques living in the Middle Atlas Mountains of Morocco. Therefore, more data on the social behaviour of wild populations are necessary before we can draw firm conclusions on the social style of this species.

11.3 The Barbary Macaque and Its Interactions with Humans: A Historic Perspective

The recorded history of the interactions between humans and Barbary macaques goes back in time to the first civilizations of the Mediterranean area (Goudsmit and Brandon-Jones 2000). At that time, the Barbary macaque was probably distributed from Morocco to Egypt, if not further East along the Mediterranean coast, and maybe also in some parts of Europe (e.g. Spain; Taub 1984). The Phoenicians, around the tenth century B.C., were probably the first civilization to report the presence of the Barbary macaque when they began trading in North Africa and across the Mediterranean Sea (Fa 1984b; Goudsmit and Brandon-Jones 2000). The Barbary macaque is also likely the tailless monkey reproduced in various Greek, Etruscan, and Italian artefacts (McDermott 1938, cited in Fa 1984b). Reports of the Barbary macaque can also be found in Aristotle's work, which classified this species as being a distinct group from the baboons known at his time. It is difficult to determine the frequency and nature of human-macaque interactions in early historic times. However, McDermott (McDermott 1938, cited in Fa 1984b) reports that some North African civilizations considered Barbary macaques solar deities. Barbary macaques were known to the Egyptians, and several remains, attributable to this species, found in Italy and the rest of Europe suggest that they were kept as pets (Groves 2006). Moreover, the Greek historian Diodorus describes three cities near Carthage (modern Libya) where the monkeys lived in the same houses as humans and shared food with them (Fa 1984b). This description closely resembles what currently happens in some villages and cities of India, where langurs and macaques live in close association with humans (e.g. Srivastava and Begum 2005). Some other North African human populations, however, were reported by Herodotus and Lucian to eat monkeys. In this respect, Fa (1984b) claims that the nature of human-macaque interactions differed between geographic regions. It is possible that tribes from the region now corresponding to Tunisia and Libya regularly ate monkeys as part of their diet, whereas this habit was not present in north-east Africa (Morocco and Algeria; Fa 1984b). Such differences may have originated from the same belief, which viewed monkeys as sacred animals (Joleaud 1931, cited in Fa 1984b), but resulted in two different traditions. According to this explanation, some tribes ate the monkeys to obtain their strength while others protected them and tolerated, if not welcomed, the presence of the monkeys in and around their cities.

11.4 Human-Macaque Interactions in Gibraltar

The free-ranging provisioned population of Barbary macaques in Gibraltar lives in the Upper Rock Nature Reserve, a protected area of around 97 ha (Shaw and Cortes 2006; Cortes and Shaw 2006). The first documented evidence of macaques in

Gibraltar dates to 1740, though speculation remains that they were present much earlier than this time (Fa 1981). Accurate historical estimates of population numbers are hard to come by (Shaw and Cortes 2006), but it is known that, from 1946 until the late 1980s, population management kept numbers between around 35 and 40 individuals (Martin 1997). After this management ceased, population size increased rapidly, reaching over 200 individuals in 1997 (Martin 1997). Since 1998, when population control measures began to be reapplied, the Gibraltar population has fluctuated between 170 and 250 individuals (Shaw and Cortes 2006; Shaw, personal communication 2011). Population size has remained relatively stable over the last few years, thanks to the birth control programme implemented by the reserve authorities (Cortes and Shaw 2006).

Barbary macaques represent a very significant tourist attraction, as most of the approximately 800,000 tourists coming to Gibraltar each year visit the nature reserve to see the famous 'Rock apes' (Fuentes 2006). Interactions between tourists and the macaques are extremely frequent (O'Leary and Fa 1993; Fuentes 2006; Shaw and Cortes 2006), not least because taxi drivers and other tour guides encourage tourists to approach the macaques or use food to lure the macaques to climb onto the tourists for photographs (Fuentes 2006). This practice leads to high levels of aggression from the macaques to tourists and results each year in a significant number of visitors receiving bites (Fa 1992; Fuentes 2006). Such episodes can attract international press coverage, with potentially negative consequences for the tourism industry and local economy of Gibraltar.

In addition to their interactions with tourists in the Upper Rock Nature Reserve, Barbary macaques in Gibraltar have a long history of moving down into the town itself and coming into contact – and conflict – with the local inhabitants (Cortes and Shaw 2006). This situation is not new: forays by the macaques into town to raid orchards and gardens were documented as early as 1875 (Fa 1981). Indeed, provisioning of the macaques up on the slopes of the rock was first introduced in 1913 to counter this problem, although this early attempt at range restriction failed abjectly, with a report some months later of animals 'taking over part of the town' (Fa 1981). Conflict in the town between macaques and Gibraltarians has been particularly problematic since the early/mid-1990s, when the macaque population started to increase dramatically. On several occasions, this has led to fissioning of troops and splinter groups attempting to establish home ranges that overlap urban areas (Shaw and Cortes 2006). The current management of the population has, however, largely been successful in minimising movements of macaques into town and reducing levels of macaque-human conflict (Fuentes 2006).

The close proximity and frequent occurrence of physical contact between Barbary macaques and humans both in the Upper Rock Nature Reserve and in the city of Gibraltar itself may increase the risk of infection and disease transmission between the animals and humans, and *vice versa* (Honess et al. 2006). For the macaques, introduction of zoonotic diseases could have catastrophic consequences (Honess et al. 2006). Recently, the Gibraltar macaque population has been screened and serologically tested for various diseases (Honess et al. 2006), and it was found that this population harbours at least two potentially zoonotic diseases – hepatitis A

(Honess et al. 2006) and simian foamy virus (Engel et al. 2008). The source of infection is not clear, and monkeys that were positive to hepatitis A did not show any sign of illness (Honess et al. 2006). So far at least, there is no evidence of any transmission of disease from macaques to humans at this site (Honess et al. 2006). Overall, transmission of diseases from non-human primates to humans is relatively rare (e.g. Huff and Barry 2003), whereas infection of monkeys from humans is more frequent (Weigler 1992). However, the potential for macaque-human and/or human-macaque disease transmission remains a real risk in Gibraltar, where human-macaque interactions are both frequent and close. Strategies to control the ranging patterns and the demography of the Barbary macaques, together with effective education programmes to control the behaviour of tourists towards the monkeys, are necessary to avoid having urbanised monkeys and to limit close interactions with humans. Moreover, periodic health checks of the population, followed by vaccination or treatment of ill individuals, seem important to maintain the Gibraltar population of macaques.

11.5 Human-Macaque Interactions in Morocco and Algeria

The distribution of the Barbary macaque reduced in historic times from the whole North African Mediterranean coast to Morocco and Algeria, which roughly corresponds to western Maghreb. Climate change may have been partially responsible for such a reduction. The geographic area inhabited by the Barbary macaques has become drier in the last 1,000 years, favouring the erosion of the mountains and the reduction of forest cover, which may ultimately have reduced the distribution of the macaques (Thirgood 1984). Human activities were also responsible for the decline of the species, considering that the North African coast has been inhabited by humans continuously for thousands of years. Although little is known about the Imazighen (or Berbers), who were present in the Maghreb as settled cultivators before the arrival of the Arabs in the seventh century A.D., the populations inhabiting or invading the Maghreb have been responsible for the continuous deforestation and hunting of the Barbary macaque to such an extent that the habitat and the macaque population have never been able to recover (Taub 1984). Deforestation significantly increased since colonialism as wood was exported for commercial use. For example, during the Fascist period, Libya experienced a mass colonisation of more than 30,000 peasants from Italy who cut down most of the remaining forests of the country (Thirgood 1984). Currently, deforestation, overgrazing by livestock (mainly sheep and goat herds) and various forms of human interference such as the illegal trade in infant macaques represent the main causes of the decline of this species in the wild (Mehlman 1989; van Lavieren 2008).

The largest population of Barbary macaques inhabits the Middle Atlas Mountains of Morocco (Camperio Ciani et al. 2005; van Lavieren and Wich 2009), but it is scattered across a number of medium to small forest patches. The Middle Atlas Mountains are often referred to as the 'castle of water', as approximately 40% of

Fig. 11.1 A young Barbary macaque on sale in Tangiers, Morocco (Courtesy and copyright: R. Troostwijk)

Morocco's fresh water comes from this region. The mixed cedar/oak forest of the central Middle Atlas Mountains is today still the home of approximately 5,000 macaques (van Lavieren and Wich 2009). This forest is unique and offers a large array of natural resources to humans, such as wood, lichens and charcoal production. However, these resources are being over-exploited, resulting in the degradation of the forest area. The large numbers of sheep and goat herds that graze in the forest year round are destroying the forest and its undergrowth and causing food competition for other species living in this habitat. Additionally, the illegal trade in infant macaques for the pet trade has become one of the greatest threats to the survival of the species (Fig. 11.1). Between 1996 and 2007, an estimated 300 infants were captured from the wild annually in the birth season and sold in the markets of Morocco to tourists (van Lavieren 2008). The macaques are smuggled to Europe where many eventually end up in sanctuaries or zoos. These animals are sometimes euthanised because of a lack of shelter space (van Lavieren 2008). The illegal trade in macaques to Europe originates mainly from Morocco. Although there have been a few unconfirmed cases deriving from Algeria, all confirmed cases of smuggled macaques originate from Morocco, as affirmed by Spanish customs and the ex-owners who bought their pets there.

A recent survey in Europe, conducted in 2011 by AAP (www.aap.nl), showed that around 75 macaques per year were offered for shelter to sanctuaries and zoos, between 2006 and 2010. This number is likely to be an underestimation, as not all macaques are reported to the authorities or sheltering organisations. Robinson and Redford (1991) calculated that the average maximum percentage sustainable off-take is around 2.5% of the total population of a primate species. According to this method, the maximum number of Barbary macaques that can be harvested in the Middle Atlas Mountains on a yearly basis, without it causing an immediate threat to this sub-population, is 125 macaques annually. However, such a figure does not take into account mortality rate or habitat fragmentation. This means that the actual off-take number may be close to or higher than the maximum number of sustainable off-take. Therefore, the risk of extinction for the sub-population of Barbary macaques in the Middle Atlas Mountains is extremely high (Kenney et al. 1995).

The Moroccan Primate Conservation Foundation (http://www.mpcfoundation.nl) has conducted extensive research and has initiated various projects in cooperation with the Moroccan authorities and national and international NGOs to combat this illegal trade, but so far, these actions have not been sufficient. Due to a lack of active law enforcement, there are still macaques openly sold throughout Morocco. Animal welfare is not a priority in Morocco, and until recently, it was still a common belief that there is an abundance of macaques in the wild. In 2009, a CITES capacity-building training was organised for the Customs Service, and this training demonstrated that there is still a long way to go before CITES regulations are correctly enforced in Morocco. Also, the absence of professional facilities to shelter confiscated macaques makes it hard for the Customs Service to enforce the laws. However, the Moroccan authorities have shown great interest in working on these issues, and new projects are planned for the future. A national action plan for the conservation of the Barbary macaque in Morocco was created in October 2011, and the implementation of this plan will hopefully have a positive outcome for the future of this species.

11.6 Case Study: Wildlife Tourism in Morocco

In January 2008, two of us (BM and MQ) established a longitudinal field project in the Middle Atlas Mountains of Morocco to study the ecology and social behaviour of the Barbary macaque (see http://barbarymacaque.blogs.lincoln.ac.uk). The study site is located in the oak and cedar forest near the city of Azrou (33° 24′N to 005° 12′W), at an altitude between 1,500 and 2,000 m above sea level. In 2009, a multi-disciplinary collaboration between three institutions (the Universities of Lincoln and Roehampton in Great Britain and the Ecole Nationale Forestiére d'Ingénieurs in Morocco) was established to study the nature of human-macaque interactions in the Middle Atlas Mountains from a biological and social anthropological perspective. Our ultimate goal is to analyse the effects of such interactions on the local communities and the Barbary macaque in an attempt to save the species while not negatively affecting the local economy. Our first study looked at how tourist pressure affects the behaviour of the Barbary macaque (Maréchal 2010; Maréchal et al. 2011). At a number of sites in the Middle Atlas Mountains, tourism focused on the Barbary macaque has begun and has significantly increased in the last few years. It has been proposed that developing this tourism may provide a powerful tool for the conservation of wild Barbary macaques (Mouna and Camperio Ciani 2006). The situation in the Atlas Mountains is not yet comparable to that in Gibraltar, as tourist numbers are much lower and there is no formal tourism 'industry' in place. However, as concerns have been raised about the impact of tourism on wild macaque populations elsewhere (Tibetan macaques in China: Berman et al. 2007, Formosan macaques in Taiwan: Hsu et al. 2009; Barbary macaques in Gibraltar: O'Leary and Fa 1993), studies on the effects of tourists' presence or their interactions with the animals are extremely important for the conservation of the species. For this reason, our aim was to explore the impact of tourism on the behaviour of adult male Barbary macaques belonging to one of the troops regularly visited by tourists (Maréchal 2010).

11.6.1 Methods

The study was conducted at the field site in the Middle Atlas Mountains, and data were collected between 21 February and 11 May 2010 on 12 male macaques belonging to the 'tourist' troop. The troop consisted at the time of the study of 51 monkeys: 12 adult and 1 subadult male, 12 adult and 1 subadult female, 18 juveniles and yearlings and 7 infants. Focal behavioural observation and scan sampling techniques (Altmann 1974) were used, with data collection starting at approximately 7:30 am and continuing until approximately 5:30 pm, 5 days per week. Scan samples every 20 min were used to collect data on the activities of the males (i.e. feeding/foraging, resting, moving, social and others, e.g. mating) and to quantify tourist presence and number. Focal samples lasting 10 min were used to quantify males' rates of grooming and aggressive behaviours, tourist proximity to focal animals and the occurrence of feeding interactions between tourists and the monkeys (where tourists gave food items by hand or threw food towards the macaques). Two focal samples were recorded for each male per day, one in the morning and one in the afternoon. A total of 180.5 h of data were collected.

We calculated the activity budgets (i.e. the percentage of scans spent in each activity during the study period) of the study animals and analysed the relationship between tourist numbers, their time spent with the monkeys, their distance to the monkeys and the daily frequency of friendly and aggressive interactions between the monkeys. We used seven proximity categories (i.e. in physical contact; 0 to $\leq$1 m; 1 to$\leq$2 m; 2 to$\leq$5 m; 5 to$\leq$10 m; 10 to$\leq$20 m; over 20 m) to analyse the effect of the distance between tourists and the monkeys.

Table 11.2 Activity budgets (% of time spent in each activity) of Barbary macaque groups in provisioned and non-provisioned conditions

Habitat (number of troops)	Feeding/ foraging	Resting	Moving	Social behaviour	Others	Reference
PR, Morocco (1)	21	46	16	11	6	This study
PR, Gibraltar (1)	9	32	38	10	11	Fa (1986)
NP, Morocco (4)	34	27	12	20	7	Majolo et al. (under review)
NP, Morocco (1)	49	14	22	11	4	Fa (1986)

PR provisioned, *NP* non-provisioned

11.6.2 Activity Budget of Provisioned and Non-provisioned Macaques

Table 11.2 provides data on the activity budgets of provisioned and non-provisioned Barbary macaques in Morocco and Gibraltar. Although direct comparisons with other studies are problematic as only adult males were observed in the current study, activity budgets differ substantially between non-provisioned and provisioned Barbary macaque groups. Under provisioned conditions, there is a decrease in feeding/foraging time and an increase in resting time. Previous studies have commented that in non-provisioned conditions, Barbary macaques generally spend more time feeding and foraging than resting (Fa 1986; Defter 1995; Ménard 2004). For this species, food resources in the wild are thought to be relatively evenly distributed and of low energetic value (Ménard 2002), and a long period of foraging and feeding is required to meet nutritional demands. The longer time spent resting by provisioned macaques could be because individuals gain their nutritional requirements in a short amount of time and/or due to the higher fat content of provisioned food that requires more time for digestion (Altmann 1992).

11.6.3 The Effect of Tourism on Grooming

The amount of time that macaque males spent grooming other group members was not related to the mean number of tourists present throughout the day (Pearson correlation: $r=0.225$, $n=47$, $p=0.129$) or to the proportion of time that tourists were present at the site (Pearson correlation: $r=0.243$, $n=47$, $p=0.100$). However, grooming time was significantly related to proximity between tourists and macaques (Spearman correlation: $r_s=0.886$, $n=7$, $p=0.019$); when tourists were in closer proximity, the monkeys spent less time grooming. These results indicate that tourists approaching the study animals too closely may interrupt their grooming interactions by either attracting them with food and/or otherwise disrupting their activity. In pygmy marmosets (*Cebuella pygmaea*), tourist presence was found to be unrelated to social grooming (de la Torre et al. 2000). In African lions (*Panthera leo*), it was observed that grooming was less frequent when tourists were present (Hayward

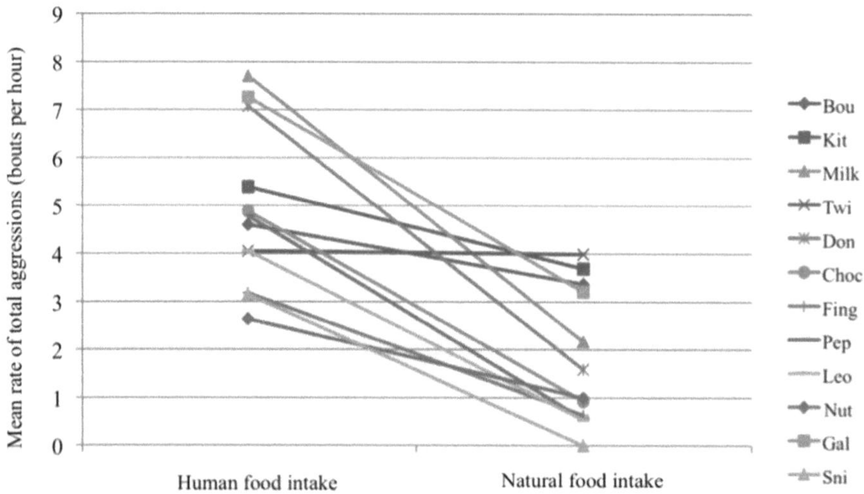

Fig. 11.2 Mean rate of aggression given or received during human and natural food intake in the study of 12 Barbary macaque males

and Hayward 2009). Bearing in mind the important hygienic and social functions of grooming (Dunbar 1991), our data suggest that close interactions between animals and tourists should be avoided.

11.6.4 The Effect of Tourism on Rates of Intraspecific Aggression

On average, males were involved in an intra-specific aggressive interaction 5.1 times per hour, a rate that is very close to that seen in a Barbary macaque group provisioned by tourists in Gibraltar (i.e. 5.2 aggression/h; Fa 1986). The rate at which males were involved in aggression interactions with other group members was not related to either the mean number of tourists present throughout the day (Pearson correlation $=0.062$, $n=47$, $p=0.677$) or to the proportion of time that tourists were present at the site (Pearson correlation: $r=0.263$, $n=7$, $p=0.074$). However, the rate of aggression was significantly associated with tourist proximity (Spearman correlation: $r_s=0.929$, $n=7$, $p=0.003$); males were more frequently involved in aggressive interactions when tourists were in closer proximity. Looking specifically at how feeding by tourists may affect aggressive behaviour, it was found that males' rate of aggression during tourist-related feeding was significantly higher than when consuming natural food (paired *t*-test, $t_{11}=6.304$, $p<0.001$; Fig. 11.2). These results suggest that the close proximity of tourists elevates rates of intraspecific aggression. This effect may again be due to the occurrence of feeding, which inevitably involves close proximity between the macaques and people. Hill (1999) compared the rates of aggression between provisioned and non-provisioned conditions in Japanese (*Macaca fuscata*) and rhesus macaques (*M. mulatta*) and found that provisioned groups had higher levels of aggression. In Tibetan macaques (*M. thibetana*), it was reported that the increase in the rate of adult aggression in provisioned monkeys was a key factor driving increased infant mortality in such troops (Berman et al. 2007). Furthermore, Hsu and colleagues (2009) recently found that food provisioning increased both the frequency and duration of aggression among Formosan macaques (*M. cyclopis*).

11.6.5 The Effect of Tourism on the Behaviour of Barbary Macaques

The results of this research are consistent with previous studies of provisioned macaque groups confronted with tourist presence. Provisioning seems to affect the activity budgets of Barbary macaques. Most importantly, the presence and behaviour of tourists affected the behaviour of the study males in ways that may have negative effects on the welfare of the animals. If tourism related to Barbary macaques in Morocco is to be developed further, it may be necessary to manage tourist sites carefully in order to eliminate or limit these negative impacts. Such mitigating measures would be welcomed as the significant economic benefits of the wildlife tourism industry could potentially make a very significant contribution to the conservation of this endangered species.

11.7 Conclusions

In this short review, we have attempted to highlight the possible problems that may arise when humans and Barbary macaques interact or come into conflict. Little is known about wild Barbary macaques or the type and effect of their interactions with humans in historic times or at present. Such information is crucial to protect this endangered species. In Gibraltar, more effective regulations should be put in place to avoid episodes of monkeys biting tourists and to limit the risk of having urbanised monkeys. Very encouraging steps have recently been taken in this direction. At the same time, education programmes could help tourists to visit the Upper Rock Nature Reserve without risk of injury. In Morocco and Algeria, more should be done to protect the Barbary macaques in their natural habitat. Strict control of illegal logging and overgrazing of livestock can have immediate benefits for the population viability of this species. In the areas that have been allocated to forest regeneration, the numbers of mixed goat/sheep herds should be minimised or the presence of such herds completely prevented. For the trade of monkeys, law enforcement against poachers, sellers and smugglers should be the highest level of priority. In order to implement effective measures to protect the Barbary macaques, the international community should work together with the Moroccan and Algerian governments. Effective controls at the borders of the Moroccan and of some European countries may significantly limit the trade in this species. Finally, the education of potential buyers of macaques is an important measure to combat the trade. In 2007, 2010 and 2011, a large-scale educational programme took place in the ports of Algeciras and Tarifa (Spain), and Tangiers (Morocco), with information handed out to the tourists on their way to Morocco for a holiday. These projects were very successful and should be repeated as often as possible. Taking away the demand for infant macaques will eventually reduce the trade and help us protect this fascinating and unusual primate species.

Acknowledgements We would like to thank Sindhu Radhakrishna, Anindya Sinha and Michael Huffman for inviting us to contribute to this volume. We thank the Haut Commissariat aux Eaux et Forêts et à la Lutte Contre la Désertification of Morocco for research permission to conduct research in Morocco. Richard McFarland and Chris Young provided useful data for this chapter and made an important contribution to our research in Morocco. Finally, we thank Abderrahmane Rakkas and Itto Fenani for their hospitality and logistical help in the field.

References

Abegg C, Thierry B (2002) Macaque evolution and dispersal in insular Southeast Asia. Biol J Linn Soc 75:555–576

Altmann J (1974) Observational study of behaviour sampling methods. Behaviour 49:227–265

Altmann SA (1992) Foraging for survival: yearling baboons in Africa. The University of Chicago Press, Chicago/London

Berman CM, Li J, Ogawa H, Ionica C, Yin H (2007) Primate tourism, range restriction and infant risk among Macaca thibetana at Mt Huangshan, China. Int J Primatol 28:1123–1141

Camperio Ciani A (1986) La *Macaca sylvanus* in Marocco: sopravvivenza o estinzione. Osservazioni personali e dati storico-demografici. Antropol Contemporanea 9:117–132

Camperio Ciani A, Palentini L, Arahou M, Martinoli L, Capiluppi C, Mouna N (2005) Population decline of *Macaca sylvanus* in the Middle Atlas of Morocco. Biol Conserv 121:635–641

Cortes J, Shaw E (2006) The Gibraltar macaques: management and future. In: Hodges JK, Cortes J (eds) The Barbary macaque: biology, management and conservation. Nottingham University Press, Nottingham, pp 199–210

de la Torre S, Snowdon CT, Bejarano M (2000) Effects of human activities on wild pygmy marmosets in Ecuadorian Amazonia. Biol Conserv 94:153–163

Deag JM (1974) A study of the social behaviour and ecology of the wild Barbary macaque *Macaca sylvanus*. PhD thesis, University of Bristol, UK

Deag JM (1977) The status of the Barbary macaque (*Macaca sylvanus*) in captivity and factors affecting its distribution in the wild. Academic, New York

Deag JM (1984) Demography of the Barbary macaque at Ain Kahla in the Moroccan Moyen Atlas. In: Fa JE (ed) The Barbary macaque: a case study in conservation. Plenum Press, New York, pp 113–133

Defter TR (1995) The time budget of a group of wild woolly monkeys (*Lagothrix lagotricha*). Int J Primatol 16:107–120

Delson E (1980) Fossil macaques, phyletic relationships and a scenario of deployment. In: Lindburg DE (ed) The macaques: studies in ecology, behavior and evolution. Van Nostrand, New York, pp 10–30

Dixson AF (1998) Primate sexuality: comparative studies of the prosimians, monkeys, apes, and human beings. Oxford University Press, Oxford

Dunbar RIM (1991) Functional significance of social grooming in primates. Folia Primatol 57:121–131

Engel GA, Pizarro M, Shaw E, Cortes J, Fuentes A, Barry P, Lerche N, Grant R, Cohn D, Jones-Engel L. (2008) Unique pattern of enzootic primate viruses in Gibraltar macaques. Emerging Infctious Diseases, 14, 1112

Fa JE (1981) The apes on the rock. Oryx, 16, 73-76.

Fa JE (1984a) The Barbary macaque: a case study in conservation. Plenum Press, New York

Fa JE (1984b) Structure and dynamics of the Barbary macaque population in Gibraltar. In: Fa JE (ed) The Barbary macaque: a case study in conservation. Plenum Press, New York, pp 3–18

Fa JE (1986) Use of time and resources by provisioned troops of monkeys. Social behaviour, time and energy in the Barbary macaque (*Macaca sylvanus*) at Gibraltar. Karger, Basel/New York

Fa JE (1989) The genus *Macaca*: a review of taxonomy and evolution. Mamm Rev 19:45–81

Fa JE (1992) Visitor-directed aggression among the Gibraltar macaques. Zoo Biol 11:43–52

Fa JE, Lind R (1996) Population management and viability of the Gibraltar Barbary macaques. In: Fa JE, Lindburg DG (eds) Evolution and ecology of macaque societies. Noyes, Park Ridge, pp 270–290

Fooden J (1976) Provisional classification and key to living species of macaques (Primates: *Macaca*). Folia Primatol 25:225–236

Fooden J (2007) Systematic review of the Barbary macaque, *Macaca sylvanus* (Linnaeus, 1758). Fieldiana Zool 113:1–60

Fuentes A (2006) Patterns and context of human-macaque interactions in Gibraltar. In: Hodges JK, Cortes J (eds) The Barbary macaque: biology, management and conservation. Nottingham University Press, Nottingham, pp 169–184

Goudsmit J, Brandon-Jones D (2000) Evidence from the baboon catacomb in North Saqqara for a west Mediterranean monkey trade route to Ptolemaic Alexandria. J Egypt Archaeol 86: 111–119

Groves C (2006) From Calpe to catanduanes: Babewynes, apes, marmesettes and othere dyverse bestes. In: Hodges JK, Cortes J (eds) The Barbary macaque: biology, management and conservation. Nottingham University Press, Nottingham, pp 1–16

Hanya G, Ménard N, Qarro M, Ibn Tattou M, Fuse M, Vallet D, Yamada A, Go M, Takafumi H, Tsujino R, Agetsuma N, Wada K (2011) Dietary adaptations of temperate primates: comparisons of Japanese and Barbary macaques. Primates 52:187–198

Hayward MW, Hayward GJ (2009) The impact of tourists on lion *Panthera leo* behaviour, stress and energetics. Acta Theriol 54:219–224

Hill DA (1999) Effects of provisioning on the social behaviour of Japanese and rhesus macaques: implications for socioecology. Primates 40:187–198

Honess PE, Pizarro M, Sene NN, Wolfensohn SE (2006) Disease transmission in the Barbary and other macaques: risks and implications for the management and conservation. In: Hodges JK, Cortes J (eds) The Barbary macaque: biology, management and conservation. Nottingham University Press, Nottingham, pp 149–168

Hsu MJ, Kao CC, Agoramoorthy G (2009) Interactions between visitors and Formosan macaques (*Macaca cyclopis*) at Shou-Shan Nature Park, Taiwan. Am J Primatol 71:214–222

Huff JL, Barry PA (2003) B-virus (Cercopithecine herpesvirus 1) infection in humans and macaques: potential for zoonotic disease. Emerg Infect Dis 9:246–250

Kenney JS, Smith JID, Starfield AM, McDouglas CW (1995) The long-term effects of tiger poaching on population viability. Conserv Biol 9:1127–1133

Lilly AA, Mehlman PT (1993) Conservation update on the Barbary macaque: declining distribution and population size in Morocco. In: A A V V (eds) Proceedings of the 16th annual meeting of the American Society of Primatologists, Sturbridge

Majolo B, McFarland R, Young C, Qarro M (under review) The effect of climatic factors on the activity budgets of a temperate primate, the Barbary macaque (*Macaca sylvanus*)

Maréchal L (2010) Tourism effects on behaviour, anxiety and physiological stress levels of wild male Barbary macaques (*Macaca sylvanus*) in Morocco. Unpublished MRes thesis, University of Roehampton, Roehampton, UK

Maréchal L, Semple S, Majolo B, Qarro M, MacLarnon A (2011) Impacts of tourism on anxiety and physiological stress in wild male Barbary macaques. Biol Conserv 144:2188–2193

Martin RD (1997) Outline proposal for effective management of the Gibraltar colony of Barbary macaques. Anthropological Institute, University of Zurich, Zurich.

McFarland R, Majolo B (2011a) Reconciliation and the costs of aggression in wild Barbary macaques (*Macaca sylvanus*): a test of the integrated hypothesis. Ethology 117:928–937

McFarland R, Majolo B (2011b) Grooming coercion and the post-conflict trading of social services in wild Barbary macaques. PLoS One. doi:10.1371/journal.pone.0026893

Mehlman PT (1989) Comparative density, demography, and ranging behavior of Barbary macaques (*Macaca sylvanus*) in marginal and prime conifer habitats. Int J Primatol 10:269–292

Ménard N (2002) Ecological plasticity of Barbary macaques (*Macaca sylvanus*). Evolut Anthropol 11:95–100

Ménard N (2004) Do ecological factors explain variation in social organization? In: Thierry B, Singh M, Kaumanns W (eds) Macaque societies: a model for the study of social organization. Cambridge University Press, Cambridge, pp 237–266

Ménard N, Vallet D (1993) Population dynamics of *Macaca sylvanus* in Algeria: an 8-year study. Am J Primatol 30:101–118

Ménard N, Vallet D, Gautier-Hion A (1985) Démographie et reproduction de *Macaca sylvanus* dans différents habitats en Algérie. Folia Primatol 44:65–81

Modolo L, Salzburger W, Martin RD (2005) Phylogeography of Barbary macaques (*Macaca sylvanus*) and the origin of the Gibraltar colony. Proc Natl Acad Sci USA 102:7392–7397

Möhle U, Heistermann M, Dittami J, Reinberg V, Hodges JK (2005) Patterns of anogenital swelling size and their endocrine correlates during ovulatory cycles and early pregnancy in free-ranging Barbary macaques (*Macaca sylvanus*) of Gibraltar. Am J Primatol 66:351–368

Mouna M, Camperio Ciani A (2006) Distribution and demography of Barbary macaque (*Macaca sylvanus*) in the wild. In: Hodges JK, Cortes J (eds) The Barbary macaque: biology, management and conservation. Nottingham University Press, Nottingham, pp 239–256

O'Leary H, Fa JE (1993) Effects of tourists on Barbary macaques (*Macaca sylvanus*) in Gibraltar. Folia Primatol 61:77–91

Robinson JG, Redford KH (1991) Sustainable harvest of neotropical forest mammals. In: Robinson JG, Redford KH (eds) Neotropical wildlife use and conservation. University of Chicago Press, Chicago, pp 415–429

Ross JF (2004) La forêt de l'Atlas menacé par les singes? Cour Int 712:24–30

Scheffrahn W, Ménard N, Vallet D, Gaci B (1993) Ecology, demography, and population genetics of Barbary macaques in Algeria. Primates 34:381–394

Shaw E, Cortes J (2006) The Gibraltar macaques: origin, history and demography. In: Hodges JK, Cortes J (eds) The Barbary macaque: biology, management and conservation. Nottingham University Press, Nottingham, pp 185–198

Srivastava A, Begum F (2005) City monkeys (*Macaca mulatta*): a study of human attitudes. In: Paterson JD, Wallis J (eds) Commensalism and conflict: the human-primate interface. American Society of Primatologists, Norman, pp 258–269

Taub DM (1975) Notes and news. Oryx 13:229

Taub DM (1977) Geographic distribution and habitat diversity of the Barbary macaque (*Macaca sylvanus L.*). Folia Primatol 27:108–133

Taub DM (1984) A brief historical account of the recent decline in geographic distribution of the Barbary macaque in North Africa. In: Fa JE (ed) The Barbary macaque: a case study in conservation. Plenum Press, New York, pp 71–78

Thierry B (2000) Covariation of conflict management patterns across macaque species. In: Aureli F, de Waal FBM (eds) Natural conflict resolution. University of California Press, Berkeley

Thierry B (2007) Unity in diversity: lessons from macaque societies. Evolut Anthropol 16:224–238
Thirgood JV (1984) The demise of the Barbary macaque habitat – past and present cover of the Maghreb. In: Fa JE (ed) The Barbary macaque: a case study in conservation. Plenum Press, New York, pp 19–69
van Lavieren E (2008) The illegal trade in Barbary macaques from Morocco and its impact on the wild population. TRAFFIC Bull 21:123–130
van Lavieren E, Wich SA (2009) Decline of the Barbary macaques *Macaca sylvanus* in the cedar forest of the Middle Atlas Mountains, Morocco. Oryx 44:133–138
von Segesser F, Ménard N, Gaci B, Martin D (1999) Genetic differentiation within and between isolated Algerian subpopulations of Barbary macaques (*Macaca sylvanus*): evidence from microsatellites. Mol Ecol 8:433–442
Weigler BJ (1992) Biology of B virus in macaque and human hosts: a review. Clin Infect Dis 14:555–567

Part V
How Living with and Beside Humans Has Affected Macaques

Utilizing women's hair as dental floss is not a simple task; the monkeys need sort the hair, make a string with it and hold it tightly with both hands to brush their teeth when they feel that pieces of food remain.

Watanabe et al.[1] *2007*

[1] Watanabe B, Urasopon N, Malaivijitnond S (2007) Long-tailed macaques use human hair as dental floss. Am J Primatol 69:940–944

Chapter 12
The Monkey in the Town's Commons, Revisited: An Anthropogenic History of the Indian Bonnet Macaque

Anindya Sinha and Kakoli Mukhopadhyay

Primas sum; primatum nihil a me alienum puto
I am a primate; nothing about primates is alien to me

Earnest Albert Hooton, 1887–1954

12.1 Introduction

"The common performing monkey of southern India" (Roonwal and Mohnot 1977), the bonnet macaque *Macaca radiata*, is remarkable for primarily two reasons – first, its inordinate ability to successfully adapt to almost any kind of environment, both natural and anthropogenic, and second, the intense love-hate relationship that it experiences with the people of peninsular India, the region of the Indian subcontinent to which it is endemic. This stems primarily from a striking feature of bonnet macaques, commented on by various authors – their inherent tendency to gravitate towards human habitations and the associated habit of becoming relatively more terrestrial rather than remaining truly arboreal (reviewed in Sinha 2001a). In this regard, the description of a monkey – most likely a bonnet macaque – "in the town's commons" in a classical Tamil poem written about 2,000 years ago (Ramanujan 1985) is truly significant.

Although many bonnet macaque populations have managed to coexist with the local people in a wide variety of non-forest habitats, including tea and coffee plantations, village agricultural areas, temples and fully urban settings (Simonds 1965;

A. Sinha (✉)
National Institute of Advanced Studies, Indian Institute of Science Campus, Bangalore 560012, India

Nature Conservation Foundation, 3076 4th Cross, Gokulam Park, Mysore 570002, India
e-mail: asinha@nias.iisc.ernet.in

K. Mukhopadhyay
44/1 Nandidurga Road, Ramya Royal Apartments, Bangalore, India

S. Radhakrishna et al. (eds.), *The Macaque Connection: Cooperation and Conflict between Humans and Macaques*, Developments in Primatology: Progress and Prospects 43, DOI 10.1007/978-1-4614-3967-7_12, © Springer Science+Business Media, LLC 2013

Schlotterhausen 1992, 1999), the intensification of agriculture in rural areas and the increasing intolerance towards the species in urban localities have led to the quiet, but steady, decline of many macaque populations, both rural and urban (Sinha 2001b; Singh and Rao 2004; Singh et al. 2011). The changes in bonnet macaque society and behaviour brought about by interactions with humans, which may be mediating such declines, however, remain unexplored. Moreover, what has not been recognised is that human-macaque interactions could also severely affect the populations of this species in our forests, some of which are protected for their wildlife. In fact, troops of completely wild bonnet macaques are now only rarely observed in interior forests over much of southern India; most troops appear to have taken up permanent residence along the roads that line and run through these forests – a significant behavioural adaptation to the tourism that is prevalent in most protected forests across the Indian subcontinent. There is, thus, an urgent need to identify the actual nature of the impact that interactions with people are wreaking on the macaques, which, in turn, may be affecting the very survival of these populations. And this is of critical importance if we are to develop conservation strategies for this and other typically commensal, but currently threatened, primate species.

Hooton, remarkable primatologist that he was, ahead of his time, could not have been more wrong!

In this chapter, we explore how the apparently simple human act of feeding monkeys, an act of great moral benevolence in the eyes of most Indians, appears to have profoundly influenced the socioecology of a population of bonnet macaques over the last decade in one of India's prime protected areas, the Bandipur National Park – Mudumalai Wildlife Sanctuary complex, in the southern part of the peninsula. These changes are significant not only because they have affected the macaques at two distinct organisational levels, those of the society and the individual, but also because they seem to be undermining the survival of the very population itself.

But we should perhaps first explain the enigmatic title of our chapter to the curious reader. A little more than 10 years ago, one of us (Sinha) had published a review of the bonnet macaque, entitled "The monkey in the town's commons: A natural history of the Indian bonnet macaque", in which he had explored the demography, socioecology and behaviour of the species as shaped by the ecological features of its largely natural, but occasionally anthropogenically modified, habitats (Sinha 2001a). We were then also beginning our quest to understand the structure and dynamics of the population of this macaque in the deciduous forests of Bandipur-Mudumalai. Today, well into this journey, we realise that the history of the species has been shaped much more by human actions than we may ever realise. The time has thus clearly come to talk of their lives and of how ours may be inexorably changing theirs forever.

12.2 The Bonnet Macaque: In Brief

Of the five primate species found in southern India, the bonnet macaque is the most common, extensively distributed from the wet rainforests of the Western Ghats mountains through the dry scrublands of central southern India to most of the hot, dusty temple towns and bustling cities of the peninsula (Krishnan 1972). The species usually live in large multimale-multifemale troops of 8–60 individuals (Sinha 2001a). Although highly variable, group sizes tend to be much smaller in natural forests and significantly larger in or near human habitations. Forest troops in Bandipur-Mudumalai, where our long-term study on the demography, ecology and behaviour of the species is in progress, for example, exhibit a mean (± SE) group size of 10.6 ± 1.2 individuals, while those inhabiting human settlements in areas adjoining these sanctuaries and around the neighbouring city of Mysore consist of 25.7 ± 3.7 individuals (Sinha et al. 2005).

Female bonnet macaques, like many other cercopithecine primates, usually remain in their natal group throughout their lives and, during adulthood, form strong, linear, matrilineal dominance hierarchies with daughters occupying dominance ranks just below those of their mothers (Sinha 2001a). Females usually develop close affiliative relationships, with high levels of allogrooming exchanged between genetic relatives as well as between unrelated individuals across the dominance hierarchy (Koyama 1973; Ali 1981). Juvenile and adult males usually emigrate from their natal troops, another typical cercopithecine feature, but bonnet macaque males appear to be unique in being rather unpredictable in this regard, some individuals even staying back to become the most dominant males in their respective natal troops (Sinha 2001a). Adult bonnet macaque males, in contrast to females, form unstable dominance hierarchies through direct aggression and coalitions, tending to occupy low ranks when very young or old but relatively high positions when mature and in the prime of their health. Bonnet macaque males are also unique among cercopithecine species in demonstrating high levels of allogrooming and other affiliative behaviour towards on another, comparable in quality and levels to that shown by the females (Ali 1981; Sinha et al. 2005). The absence of any correlation between individual dominance ranks and the levels of such affiliative behaviour displayed or received indicates that bonnet macaque males may, in general, adopt individual social strategies that are much less constrained by the rank hierarchy than is typical for males of most cercopithecine species living in multimale groups (Sinha et al. 2005).

In terms of sexual behaviour, bonnet macaques are typically promiscuous with ample mating opportunities for both males and females. Adult males appear to follow different sexual strategies, with some males forming consortships of varying periods of time, while others simply copulate opportunistically with the available females (Simonds 1965; Sugiyama 1971; Sinha 2001a). Whatever be the strategy, which varies across individuals and situations, only rarely do bonnet males interrupt the copulation of others or harass them in any way, even if they are subordinate males (Ali 1984; Sinha 2001a). Subadult or juvenile males are also not denied access to oestrus females. Finally, although bonnet macaque females appear to be remarkably undiscriminating, mating freely with males of all ages and ranks (Simonds 1965; Sugiyama 1971), females often actively exercise mate choice by rather subtle manipulative strategies (Sinha 2001a). Bonnet macaque females also clearly prefer to sexually solicit new immigrant males over resident males, even if the latter are much more dominant (Sinha 2001a).

12.3 The Population: In Focus

The study population primarily consists of a total of 30-odd troops (some of which have disappeared or died out and been replaced by new immigrant troops over the study period of 11 years) inhabiting the southeastern region of Bandipur National Park in Karnataka state and a small adjoining section of the contiguous Mudumalai Wildlife Sanctuary in Tamil Nadu state (*c.* 11°57′–11°76′N, 76°55′–76°68′E). Although a range of tropical forest vegetation types – from moist deciduous forest through semi-evergreen forest to dry evergreen forest – occur within these sanctuaries, the study groups largely inhabit areas with predominantly dry and moist deciduous vegetation. The demographic structure of some of these troops has been described elsewhere for the interested reader (Sinha et al. 2005).

The natural diet of the study troops consists mainly of fruits, flowers or leaves of *Tamarindus indica*, *Azadirachta indica*, *Ficus benghalensis*, *Ficus retusa*, *Alphonsea sclerocarpa*, *Zizyphus oenoplia*, *Memecylon edule*, *Randia malabarica*, assorted herbs, as well as tender shoots of several grasses, including bamboo (*Bambusa arundinacea*), particularly in the riverine tracts of the study area. Insects, including crickets and grasshoppers, also contributed to their diet. More importantly and of greater relevance to this study, some of the troops occasionally feed on high-calorie human food handed out by tourists passing along the highway running through the sanctuaries. Free-ranging groups of bonnet macaques and occasionally common langurs, *Semnopithecus entellus*, are typically provisioned by tourists visiting a number of wildlife sanctuaries in southern India. The food obtained during such interactions is nutritionally rich but patchy and clumped in distribution. The amount of such food specifically available to the study troops is, however, seasonal, unpredictable and directly proportional to the tourist inflow into the sanctuaries. For some troops, therefore, such food forms only a small fraction of their total dietary intake. These troops thus regularly forage on their natural food sources and only resort to provisioned food during particular seasons and during certain times of the day when tourist traffic peaks within the sanctuaries. Several troops in this population are also substantially provisioned owing to their home ranges being around human habitations (of the Forest Department) and feed daily on human foods discarded by the local people.

Our analysis of the impact of human provisioning on the social organisation and behaviour of the macaque population, presented in this chapter, is based on data acquired through the demographic monitoring of all the troops, conducted every 3–4 months over the last 11 years (beginning March 2000), and structured behavioural observations on identified individuals in select troops, conducted in two phases, between December 1996 and March 1997 and between May 2005 and June 2008.

12.4 Usual Population, Unusual Troops

Our first arrival in Bandipur-Mudumalai in the spring of 2000 was greeted by a surprising discovery: 11 of the 21 troops that we initially surveyed were unimale-multifemale; each of these troops consisted of a single adult male, accompanied by a few adult females and their dependent young, both infants and juveniles (Sinha et al. 2005). This was surprising because bonnet macaques have always been reported to reside in multimale-multifemale groups, a feature common to most cercopithecine primates (Roonwal and Mohnot 1977; Smuts et al. 1987). What was also noteworthy is that several earlier demographic studies of this species in this particular area had either not explicitly reported the occurrence of unimale troops (Singh et al. 1984; D'Souza and Singh 1992) or had provided evidence of only a very small proportion of such groups in the population – 12.5–13.8% (Simonds 1965; Kurup 1981), as compared to our fairly significant value of 52.4% (Sinha et al. 2005).

The unimale troops, with a mean (± SE) of 10.82 (± 0.98) individuals in each group, were significantly smaller than the multimale troops, each with a mean of 20.20 (± 2.97) individuals and had less number of adult females (Sinha et al. 2005). In contrast to earlier reports (Sugiyama 1971; Simonds 1974), the adult male/adult female ratio exhibited striking variation – ranging from 0.25 to 1.0 for the unimale groups and from 0.60 to 2.0 for the multimale troops (Sinha et al. 2005).

What factors, ecological or biological, could have led to this unprecedented, dramatic increase in the proportion of unimale social groups, a rather unusual form of social organisation in bonnet macaques, in this population in recent years?

12.5 An Ecological Drive in Social Evolution

The adult sex ratio in most primate groups, including macaques, appears to be related to the length of the breeding season and the degree of oestrus synchrony among females (Ridley 1986; Oi 1996; Nunn 1999). Bonnet macaques generally live in seasonal environments, and accordingly, most females within a troop come into oestrus synchronously. As compared to the situation in certain other species such as the lion-tailed macaque (Kumar and Kurup 1985) or the pig-tailed macaque (Oi 1996), this seems to have led to a relatively greater proportion of males within natural groups of bonnet macaques and correspondingly a promiscuous mating system. Is it, however, possible that adult females within unimale bonnet macaque troops have a lower degree of oestrus synchrony (as prevails among species inhabiting non-seasonal environments such as pig-tailed or lion-tailed macaques) enabling a single adult male to monopolise each female as they independently become sexually receptive? Representative demographic data, shown only for the initial years of this study, indicate a comparable temporal clustering of births in both unimale and multimale troops (Fig. 12.1); this would, thus, argue against the causal mechanism described above for the evolution of this form of social organisation.

An ecological factor that may have significantly affected primate communities in Indian wildlife sanctuaries in the recent past is the increasing tourist traffic within these protected areas. As bonnet macaques appear to naturally gravitate towards humans and their food sources, many tourists invariably come in contact with such macaque troops and provision them with human foods. Such food is often nutritionally rich, and provisioning is thus marked by intense within-troop contest competition and scramble competition among neighbouring troops, especially along the

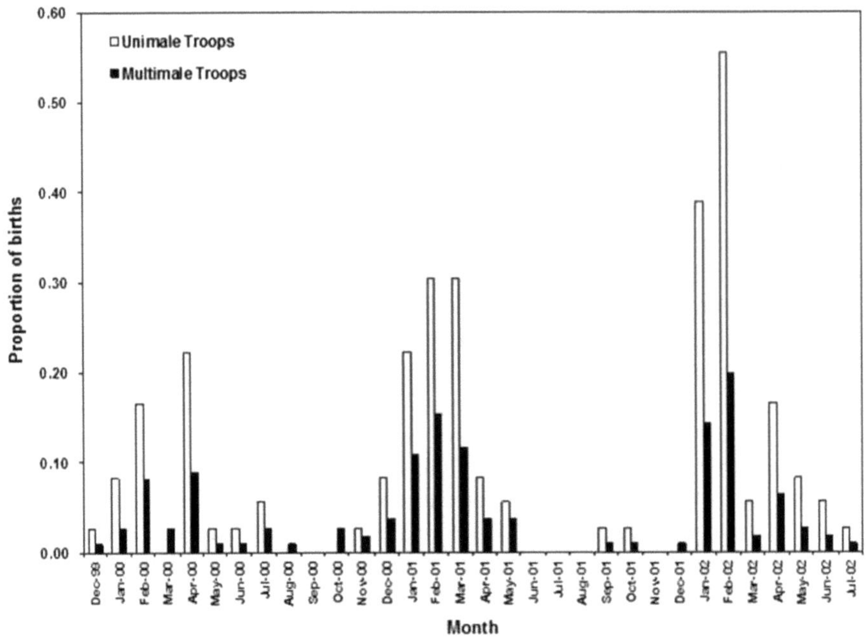

Fig. 12.1 Temporal clustering of births in unimale and multimale troops of bonnet macaque in the Bandipur-Mudumalai population. Representative demographic data are shown for 11 unimale (36 infants) and 14 multimale (111 infants) troops during the period from December 1999 to July 2002

highway (Mukhopadhyay and Sinha, unpublished data; see also van Schaik 1989). This is particularly accentuated during the summer months when natural food resources within the dry deciduous forests are sparse and patchily distributed, and most of the troops come and space themselves along the highway (Mukhopadhyay and Sinha, unpublished data). The amount of provisioned human food, however, is also rather unpredictable and clumped in distribution and, like the natural food sources available at this time, can support possibly only small groups of macaques. Bonnet macaque females, being philopatric, form stable core groups in a particular area, and this may have led to the evolution of unusually small troops of closely related females in this population. It is likely that such groups are then easily reproductively monopolised by a single dominant adult male, as has been postulated earlier for other species from a comparative analysis of cercopithecine mating systems (Smuts et al. 1987). Our model, depicting such an evolutionary pathway for unimale social organisation among bonnet macaques in the Bandipur-Mudumalai population, is shown in Fig. 12.2.

Support for such an anthropogenic pressure possibly driving the relatively rapid evolution of unimale troops in this population in recent years comes from four different sources. First, we have now documented at least ten instances of troop fission (in 7 of the 30 troops observed between March 2000 and September 2011)

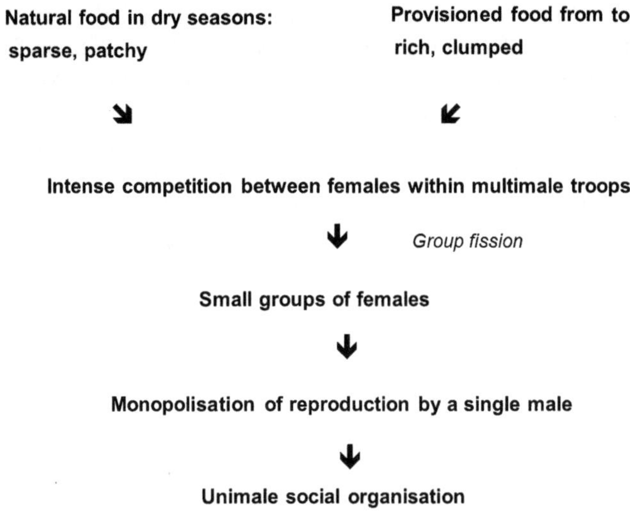

Fig. 12.2 A model depicting the possible evolutionary route for unimale social organisation in the bonnet macaque population of Bandipur-Mudumalai (Adapted from Sinha et al. 2005)

during which several females (ranging in number from two to four females, constituting approximately 25–60% of the resident adult females in a troop) left their natal troops and emigrated, occasionally in the company of one or more adult males. Such a high rate of group fission among females has rarely been reported in any female-bonded species.

A second line of evidence stems from our observation that in Bandipur-Mudumalai and their surrounding areas, bonnet macaque troops that inhabit human habitations are able to amply forage on unlimited sources of human food; such troops are invariably multimale and occupy relatively large home ranges. Thus, of the 21 troops described during the initial part of our study (Sinha et al. 2005), 7 were observed to live in the vicinity of human settlements, and only one of these troops was unimale in composition. In contrast, 10 of the 14 genuinely forest troops were unimale, the selectivity of habitats being significantly different across the two types of troops. Interestingly, as has been predicted for competition between groups for rich, but patchily distributed, food resources (Wrangham 1980; van Schaik 1989), troops living near human habitations were larger than the forest troops (mean group size ± SE of 21.9 ± 3.1 for habitation troops and 10.6 ± 1.3 for forest troops) while the distribution of their troop sizes was also significantly different (Sinha et al. 2005). Moreover, an extensive survey of bonnet macaque populations within these sanctuaries, conducted in April 2005 and in October 2008, revealed only very few or no unimale troops along other highways that were not as heavily frequented by tourists (Mukhopadhyay and Sinha, unpublished data).

The final two lines of evidence in support of our model come from our behavioural studies, conducted between May 2005 and June 2008, on four – two multimale and two unimale – troops of macaques in the Bandipur National Park. One troop of each type (MM 1 and UM 1) occurred close to a human habitation (Bandipur Camp) while the two other troops (MM 2 and UM 2) occurred in a forested area bordering the highway, approximately 2 km away.

First, the two forest troops, UM 2 and MM 2, had significantly larger home ranges during the wet season (25.01 and 36.49 ha, respectively) than did the two troops close to the human habitation (170.95 and 236.06 ha, respectively; Chatterjee et al. 2009). This was possibly due to the ease with which the two latter troops were able to access anthropogenic food resources, particularly those that originated from visiting tourists. Competition over food was enhanced during the dry summer months when the home range size increased significantly – but only for the two multimale troops (MM 1: 36.49–235.10 ha and MM 2: 236.06–268.91 ha, in contrast to UM 1: 25.01–27.46 ha and UM 2: 170.95–120.39 ha; Chatterjee et al. 2009). This is thus a clear indication that relatively larger troops need to increase their home range when food resources become limiting (see also Isbell 1991).

Finally, we examined the frequency and nature of competitive inter-group interactions between these macaque troops in their natural and anthropogenic habitats and across the two seasons that distinctly differed in food availability (Chatterjee et al. 2009). Both scramble competition (in which one troop retreated from another advancing troop without any physical interaction between the members of the two troops; van Schaik 1989) and contest competition (in which there were agonistic behavioural interactions between the members of the two troops) were displayed by the two provisioned troops, UM 1 and MM 1, at levels two to three times higher than that shown by the two forest troops. The provisioned troops also tended to initiate such interactions more during the resource-strapped dry season though the observed differences were not statistically significant (Chatterjee et al. 2009). Moreover, the larger multimale troops tended to be involved in a relatively higher proportion of inter-group interactions and also initiated such interactions at a significantly higher frequency.

12.6 Unimale and Multimale Troops: The Rise of the Despot!

The largely anthropogenically driven generation of unimale troops in the Bandipur-Mudumalai population of bonnet macaques appears to closely relate to the appearance of striking behavioural differences between the single male of a unimale troop and the alpha (or the most dominant) male of a multimale troop (Sinha et al. 2005).

In multimale bonnet macaque troops, the adult males are remarkably tolerant of the reproductive efforts and acts of other adult and subadult males (Simonds 1965; Sugiyama 1971; Sinha 2001a). In unimale troops, however, the resident male enjoys a complete reproductive monopolisation of the troop females, and this appears to be related to a strong intolerance of other males, both within and outside the group.

As prevails in most unimale harem systems (Smuts et al. 1987), there are several behavioural manifestations of this intolerance, each of which appears to be unique to unimale troops and which have virtually never been encountered within multimale social groups in any bonnet macaque population that we have studied so far (elaborated in Sinha et al. 2005). The most notable of these behavioural differences includes the display of severe aggression by the solitary adult male of a unimale troop towards the subadult and juvenile males in his troop. Such aggression is usually much lower or less unidirectional within multimale troops.

The resident adult male in all unimale troops typically herds the group females; such herding usually intensifies in the presence of neighbouring troops, especially when extra-troop males approach the unimale group. Typical of certain other unimale primate social organisations (Smuts et al. 1987), regular herding of this kind has never been seen in multimale troops in any population except during periodic consortships when certain dominant males intensely guard particular oestrus females one-on-one for varying periods of time, successfully preventing other intra-troop males any access to them (Sinha 2001a). The resident male of unimale troops invariably defends the troop during inter-troop encounters and physically interacts with the male(s) of the other troop, sometimes inflicting grievous injuries on them. In striking contrast, the most dominant males of multimale troops virtually never ever participate in inter-troop encounters, but, accompanied by the troop females, observe the interactions between the other group males and the challenging troop from a certain distance.

The adult males of unimale troops are rather successful in preventing immigration of other males into their troops as the cost that would ensue in terms of reduced future reproductive success is obviously higher than what would accrue to resident males in multimale troops with less-skewed adult sex ratios (Sinha et al. 2005). We thus observed a complete absence of any immigration into seven unimale troops that were monitored during the mating season during the period from 2000 to 2005 while a total of ten males had successfully joined the six multimale troops during the same period (Mann-Whitney U-test, $p<0.02$). There was also a strong positive correlation between the number of resident males in a troop and the number of immigrant males joining such a troop (Kendall's rank correlation, $n=13$ troops, $p<0.0001$). Resident males in large multimale troops may thus be indifferent to the entry of new males into the group because, with increase in troop size, such immigration would have progressively decreasing effect on the adult sex ratio (and hence, the per capita number of females available to the average male) in these troops.

The coexistence of two distinct kinds of societies within the same population of bonnet macaques raises several important issues. These concern, in particular, the stability of the unimale troops which must face enormous pressure from potentially immigrant males, resource competition between the smaller unimale and the larger multimale troops, patterns of male emigration and reproductive success across these groups, and the competition for food and mates that females in multimale troops face as well as the lack of mate choice that females in unimale troops must cope with.

An illuminating example of how individual bonnet macaques might be able to respond to newly emerging socioecological problems through the adoption and/or evolution of novel behavioural strategies is that of the extensive dispersal displayed by adult, subadult and juvenile females in our study population; such migration is independent of that shown by females during the events of troop fission described above. Thus, a total of 14 females, including 6 adults, 2 subadults and 6 juveniles, emigrated out of their troops, either singly or in small associations, during the first 5 years of our study (2000–2005; Sinha et al. 2005). Although dispersal by female bonnet macaques has been postulated (Ali 1981) and once reported earlier under unusual circumstances (Singh et al. 2006), the nature of female dispersal in our study population appears to be an unusual phenomenon that may have arisen in response to a lack of mate choice faced by females sometimes confined to a unimale form of social organisation. This is suggested by the fact that such widespread dispersal by female bonnet macaques of all age groups has only been observed in this particular population and never encountered within the two other populations, both close to Bangalore city (GKVK Campus, with two multimale troops and Bannerghata National Park, with a multimale and an unimale troop) that had earlier been closely observed for a combined period of more than 6 years (March 1993–December 1999). Female emigration in the Bandipur-Mudumalai population occurred in five unimale and three multimale troops with three of the six adults, both the subadults and four of the six juveniles dispersing from the former. In terms of the entire population, dispersal was thus displayed by 9 of the 62 females in unimale and 5 of the 83 females in multimale troops (G test of independence; $0.20>p>0.10$) – a trend towards relatively more emigration out of unimale troops, as would be predicted by the above hypothesis.

12.7 Ecology Proposes, Behaviour Disposes

Provisioning of wild primate troops, in addition to bringing about long-term, demographic and social changes at the group level, as outlined above, may also lead to short-term alterations in the behavioural activities and social strategies of individuals. Several studies, for example, have examined modified food choice and foraging patterns of individual macaques in response to anthropogenic habitat changes (see Altmann and Muruthi 1988; Riley 2007, 2008) or even the nature of aggression directed by macaques towards humans (such as tourists) who provision them (see O'Leary and Fa 1993; Wheatley and Harya Putra 1994; Hill 1999; Fuentes and Gamerl 2005; Fuentes 2006; Hsu et al. 2009; McCarthy et al. 2009; Chauhan and Pirta 2010).

Observations on the social interactions of adult macaques, both male and female, in response to provisioning and the nature of changing social relationships within such groups have, however, remained surprisingly neglected (see Hill 1999; Ram et al. 2003; Sinha et al. 2005; Berman et al. 2007; Hsu et al. 2009; Chap. 11 by Majolo et al. this book). The most evident behavioural response of macaques to provisioning, documented by the above studies, is a marked increase in the frequency of aggressive interactions and feeding supplants within the troop. In a detailed study on Taiwanese macaques at a nature reserve in Taiwan, for example, Hsu et al. (2009) documented an almost fivefold increase in within-troop aggression as the study group moved from natural foraging to being provisioned by tourists.

Our first studies on the bonnet macaques of the Bandipur-Mudumalai population, conducted as early as between December 1996 and March 1997, in fact, examined the immediate behavioural responses of bonnet macaques within a particular troop (BM 19) in the Mudumalai Wildlife Sanctuary and the social strategies adopted by both adult females and males under these conditions (Ram et al. 2003; Sinha et al. 2005). What was unique to this particular troop was that it regularly alternated between regimes of natural foraging and provisioning by tourists, occasionally even within the course of a single day. We summarise below our understanding of the behavioural strategies adopted by the adult females alone because, being members of a female-bonded species, the philopatric females typically live their lives together and form the core group of individuals that give a troop its characteristic behavioural patterns and its own unique identity (Sinha 2001a).

12.8 Paradise Lost...

There was a marked increase in social tension in the study group during periods of scavenging, particularly when they interacted with tourists; this was usually manifested as enhanced aggression among the relatively more dominant adult females of the troop. Thus, the overall rates of total aggression and its two components – non-contact aggression and contact aggression – as well as aggressive approach and feeding supplant increased significantly, about two- to threefold, during this period from that exhibited during natural foraging (Ram et al. 2003).

Adult bonnet macaque females typically do not exhibit aggression towards those higher than them in the dominance hierarchy (Sinha, unpublished data) while dominant individuals typically direct aggression towards those subordinates from whom retaliation is least likely. During scavenging on human foods, however, when there was a rise in overall aggression within the troop, females displayed significantly enhanced aggression towards their dominant counterparts; subordinate individuals were now likely to retaliate against the aggression that they received. It has been suggested that stable linear hierarchies in female primates are a result of conflict competition for food (van Noordwijk and van Schaik 1987; Janson and van Schaik 1988; Gore 1993). Remarkably, however, the dominance hierarchy in this particular troop of bonnet macaques became comparatively unstable under conditions of provisioning.

Potentially expensive acts of aggression over provisioned food appeared to be demonstrated more by high-ranked females, who could obviously afford such encounters, and were preferentially directed towards high-ranked subordinates who, by virtue of their position in the hierarchy, would be the most threatening adversaries during such feeding competition. Thus, challenges during contest competition would most likely come not only from individuals of high rank but from those most closely ranked in the dominance hierarchy as well.

Another mechanism that may yield a similar pattern of interactions is that low-ranking subordinate females may avoid costly conflicts by physically occupying positions away from high-ranking individuals, thus giving rise to specific group spatial structures during feeding competition (see, Barton 1993). The spatial unpredictability of the provisioned food, discussed above, however, led to a situation when all the adult females in the group scavenged together, in close proximity, and no distinct group spatial organisation could be discerned. Moreover, females across the rank hierarchy enjoyed comparable scavenging success when provisioned, in contrast to what has earlier been observed in female Japanese macaques (Imanishi 1957) and olive baboons (Barton 1993). Finally, the hypothesis that aggression may be preferentially directed towards certain individuals during enhanced feeding competition and may not simply be an emergent property of individual positional choices comes from the observation that, when provisioned, subordinate females displayed enhanced aggression towards their dominant adversaries and significantly reciprocated the aggression that they received.

There was also a significant decrease in the frequency of allogrooming and other affiliative behaviours displayed by the females as they moved to the provisioning site from areas of natural foraging (Ram et al. 2003). Relatively subordinate females, however, did not reduce the display of any of these affiliative behaviours towards their dominant partners, and their levels remained comparable across the two ecological situations. Finally, it is perhaps noteworthy that while only the relatively less time- and energy-demanding affiliative behaviours were reciprocally exchanged between the adult females of the study troop during foraging, the more demanding allogrooming was also reciprocated during provisioning (Ram et al. 2003).

It was also of great importance to examine the behavioural strategies particularly adopted by the bonnet macaques within multimale and unimale troops in order to cope with the long-term spatial and seasonal variation in the amount and pattern of food distribution in their habitat, as documented in the earlier part of this chapter. Would such an investigation provide evidence to support the model (depicted in Fig. 12.2) that attempts to explain the origin of unimale troops within our study population? Here, we discuss briefly the social behaviour of the largely philopatric adult females in the two multimale (MM 1 and MM 2) and the two unimale (UM 1 and UM 2) troops, described above, as the model postulates their behavioural strategies to be instrumental in triggering of the social changes that may have led to the formation of this unique form of social organisation within the Bandipur-Mudumalai population of this female-bonded cercopithecine primate.

Of the four study troops, the two troops in the forested tracts, MM 2 and UM 2, spent significantly more time feeding on natural food resources than on those provisioned by tourists (Wilcoxon's matched-pairs signed-ranks test; MM 2: $T=45$, $n=9$, $p=0.008$; UM 2: $T=21$, $n=3$, $p=0.026$). Close to human habitations, however, only the unimale troop UM 1 spent a significantly higher proportion of time on natural foraging; the multimale troop MMI, in contrast, spent comparable proportions of time foraging on both kinds of food resources (UM 1: $T=21$, $n=6$, $p=0.014$; MM 1: $T=32$, $n=9$, $p=0.26$).

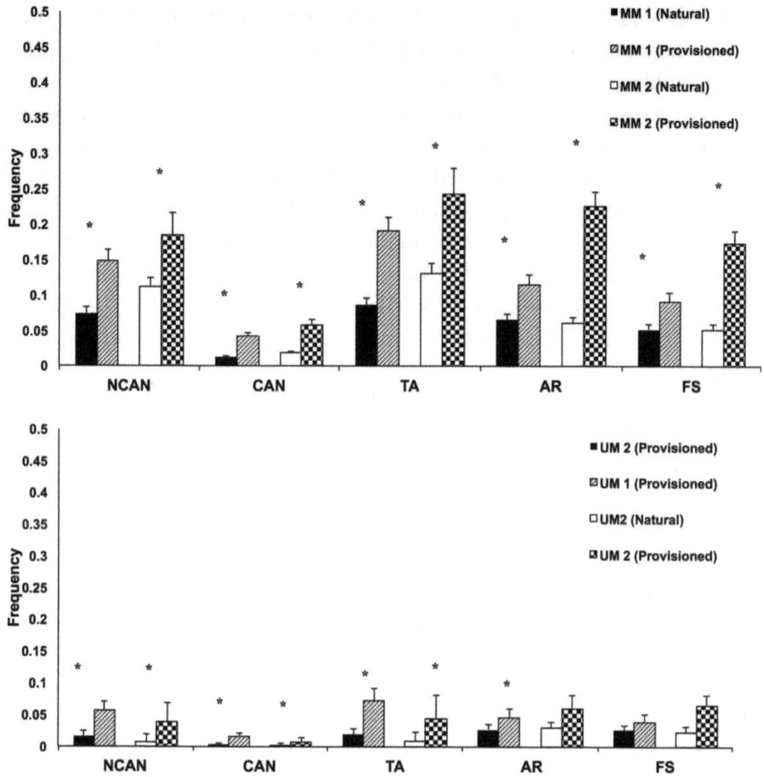

Fig. 12.3 Aggressive behaviours displayed by the study troops during foraging and feeding on natural and provisioned resources. (**a**) Multimale troops MM 1 and MM 2; (**b**) Unimale troops UM 1 and UM 2. The *bars* represent mean frequency and standard error of behavioural acts/individual/h (ordinate) for the different behaviours while the *asterisks* indicate a statistically significant difference between each respective pair of values (matched-pairs test, $p < 0.05$; see Ram et al. (2003) for details of the test). *NCAN* non-contact aggression, *CAN* contact aggression, *TA* total aggression, *AR* aggressive approach, *FS* feeding supplant (see Ram et al. (2003) for description of the behaviours)

In general, the females of the relatively larger multimale troops displayed significantly higher levels of aggression, including non-contact aggression, contact aggression, aggressive approach and feeding supplants, than did females in the smaller unimale troops (Chatterjee et al. 2009). Significantly, these results also held true when analysed separately for both feeding regimes and both seasonal phases, except for under conditions of natural foraging in the wet season.

The adult females in all four troops, however, exhibited all four forms of aggression at significantly higher frequency when they were provisioned as compared to when they fed on natural resources (Fig. 12.3). It is perhaps noteworthy that the frequency of feeding supplants was comparatively low for the females in the two unimale troops and did not differ significantly across the feeding regimes; there was also no significant difference in the frequency of aggressive approach displayed by the females of troop UM 2 under these conditions (Fig. 12.3b).

There was, however, a significant change in the time spent foraging on natural and provisioned resources across seasons, with troop MM 1 spending a greater proportion of time foraging and feeding on provisioned food in the dry season (Wilcoxon's matched-pairs signed-ranks test; $T=5$, $n=9$, $p=0.04$) while the other three troops primarily foraged on natural resources. In the wet season, both the multimale troops and the unimale troop UM 2 preferentially indulged in natural foraging (MM 1: $T=40$, $n=9$, $p=0.04$; MM 2: $T=36$, $n=9$, $p=0.012$; UM 2: $T=21$, $n=3$, $p=0.03$); troop UM 1, however, spent comparable time on both kinds of food resources.

There was a significant escalation in all forms of aggression related to feeding competition among the study females in both multimale troops from the wet to the dry season (Fig. 12.4a); in contrast and most importantly, there were no such seasonal differences in the displayed frequency of any of the aggressive behaviours in either of the two unimale troops (Fig. 12.4b).

Finally, an examination of individual behavioural profiles in the study troops revealed that almost all the aggression displayed by adult females during foraging and feeding was directed significantly towards their subordinate counterparts; aggression directed up the hierarchy was negligible (Chatterjee et al. 2009). These patterns were consistent for all the troops under the different feeding regimes and across the two seasons.

As food supplants directly measure the effect of food competition in hindering individuals from obtaining food resources, we also investigated the distribution of feeding supplants, with respect to dominance status, in all troops under the different seasonal conditions and feeding regimes (Chatterjee et al. 2009). There were no significant patterns in the two unimale troops under all conditions and in the two multimale troops in the wet season though a significant relationship was revealed in the latter troops during the competitive dry season. In the multimale troop MM 1 that inhabited Bandipur camp, the females lowest in the dominance hierarchy were supplanted less often than were subordinate females relatively higher up in the hierarchy. In the other multimale troop MM 2 that resided in a forest area, however, relatively low-ranked females were supplanted significantly more than were those of higher rank. These patterns held true for both troops during natural foraging and on provisioning by tourists although the relationship was statistically significant only under the latter regime (Chatterjee et al. 2009).

Our observations thus not only confirm the critical function of food resources in governing female social relationships in bonnet macaques, as may have been expected, but also underscore the role of behavioural flexibility in the species that permits individuals to adjust their behaviour to the prevailing ecological conditions. The direct effect of group size on within-group contest competition is revealed by the elevated levels of food-related aggression displayed by the females in large multimale troops alone and particularly so during the dry season. A scarcity of food resources, therefore, does induce a high degree of contest competition among female

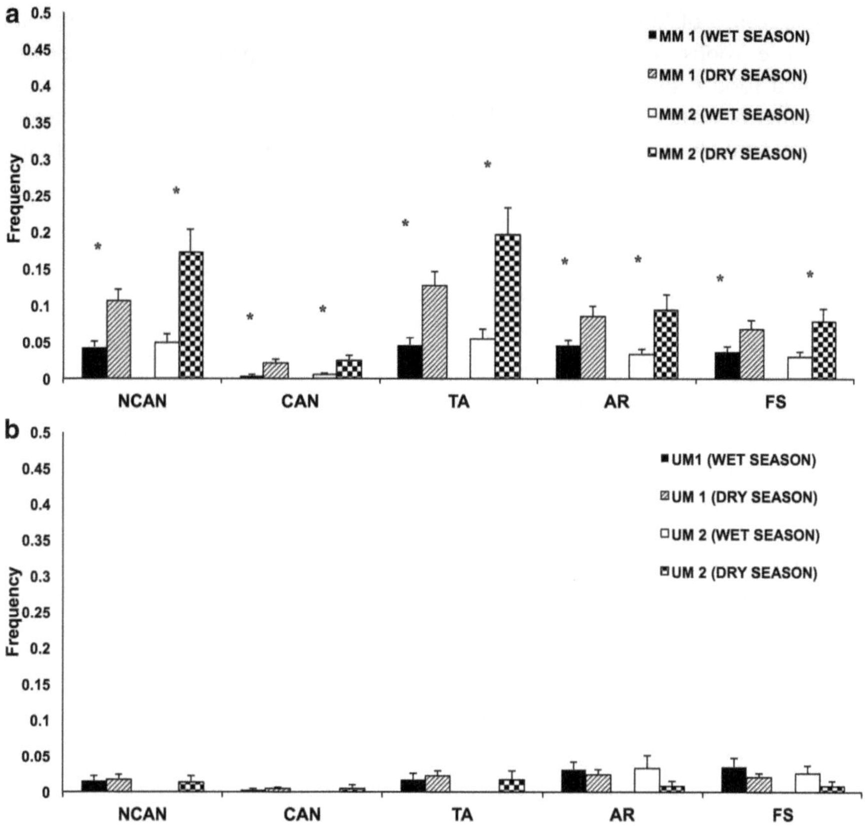

Fig. 12.4 Aggressive behaviours displayed by the study troops during foraging and feeding in the wet and dry seasons. (**a**) Multimale troops MM 1 and MM 2; (**b**) Unimale troops UM 1 and UM 2. The *bars* represent mean frequency and standard error of behavioural acts/individual/h (ordinate) for the different behaviours while the *asterisks* indicate a statistically significant difference between each respective pair of values (matched-pairs test, $p < 0.05$; see Ram et al. (2003) for details of the test). *NCAN* non-contact aggression, *CAN* contact aggression, *TA* total aggression, *AR* aggressive approach, *FS* feeding supplant (see Ram et al. (2003) for description of the behaviours)

bonnet macaques but only when they are in large aggregations. Several authors have earlier noted that provisioning dramatically increases contest competition among female primates (Southwick et al. 1976; Mori 1977; Hill 1999; Ram et al. 2003). The increase in aggression among females over such provisioned food directly relates to the nature of the food resource – provisioned human food is usually richer in calorific value than are typical natural food resources, relatively more clumped in distribution and usually temporally unpredictable in their availability (Hill 1999; Saj et al. 1999). Our results thus appear to support our model (Fig. 12.2) that female social relationships fuelled by food competition may be primarily responsible for the rise of the unusual unimale form of social organisation in the Bandipur-Mudumalai population of bonnet macaques. We have also provided evidence above of how groups of low-ranked subordinate females may decide to leave their natal multimale troops, possibly in the face of increasing competition over scarce and clumped food resources. It should, however, be pointed out here that our final prediction of certain adult males taking over the resulting small groups of females to form unimale troops has not been explicitly tested as yet and requires future confirmation.

12.9 ...And Paradise Regained!

While studies such as ours and those listed above are essential for a basic understanding of the behavioural strategies that individuals display when faced with rapidly changing food regimes, there is also an urgent need to explore the mechanisms that facilitate the promotion of social harmony in the midst of rising intra-group competition for food, often of human origin.

Our study of the Mudumalai troop, BM19, conducted between December 1996 and March 1997, thus, also attempted to understand the processes by which individual bonnet macaques were potentially able to lessen social tension under the difficult conditions of extreme feeding competition (Ram et al. 2003; Sinha et al. 2005). There appeared to be a marked tendency for the adult females in the study troop to avoid one another as they moved from a regime of natural foraging to that in which they scavenged aggressively for provisioned food (Ram et al. 2003). This was evident from the sharp increase in the frequency of retreat displayed by all the females, irrespective of their dominance ranks, while scavenging. Surprisingly, there was a significant positive correlation between the frequency with which females retreated from their dominant adversaries and the absolute dominance rank of the retreating individuals, but only when the troop was provisioned – higher-ranked individuals thus tended to retreat more under more competitive regimes of resource availability.

Among the affiliative behaviours, allogrooming was clearly exchanged between females in the group as a general strategy only under conditions of scavenging, as noted above. In addition, subordinate females allogroomed their dominant partners significantly more than were reciprocated, when they were provisioned; the levels of grooming exchanged were much more comparable during natural foraging.

Finally, the females were observed to particularly allogroom and express higher levels of other affiliative behaviours towards those dominant adversaries who displayed relatively higher levels of aggression towards them or approached them aggressively during provisioning (Ram et al. 2003). Such behavioural interchange was, however, strikingly absent during social interactions associated with natural foraging. These examples of specific behavioural exchanges clearly seem to suggest a strategic role for certain kinds of affiliative behaviour, particularly allogrooming, in reducing social tensions that prevail within a troop during particularly competitive regimes.

The adult females in the study group, therefore, appeared to exhibit two distinctive kinds of behavioural strategies to reduce social tension brought about by enhanced feeding competition. First, individuals of high dominance rank actively avoided interacting with more dominant females, actively retreating from them at much higher rates than they did under the less stressful conditions of natural foraging. Avoidance of this kind may be the only strategy that can be utilised by individuals of high dominance rank to alleviate intra-group aggression. Given the tension that individuals seem to be under during provisioning, attempts to bring about peace through more active affiliative interactions could carry the threat of potentially high costs of physical conflict, a cost that would be maximal for individuals of high dominance rank interacting with one another.

The second kind of behavioural strategies consists of active affiliative interactions directed towards those individuals from whom the threat of aggression is the most or towards those who are victims of contest competition. This was originally suggested by de Waal (1989) who emphasised that primates, being highly adaptable, would, when confronted with a stressful situation, modify their behaviour to reduce the risk of aggression.

12.10 Food for Thought

The clumped distribution of the provisioned food around their human sources was likely to be directly responsible for the observed significant increase in intra-group aggression, aggressive approaches and feeding supplants displayed by our study troop during these periods over that during natural foraging. Bonnet macaques are generalist feeders (Sinha 2001a) and usually space themselves out as they forage on widely dispersed, abundant leaves, flowers and fruits of several commonly available plant species. Opportunities for monopolisation of food are, therefore, rare during natural foraging and strong competition unlikely to occur under these circumstances.

The provisioned food, in contrast, was markedly clumped in distribution, not only in time, as noted above, but also in space – within a small area habituated by large numbers of tourists. Although concentrated within this particular area, the most important feature of such food, nevertheless, was its exact spatial unpredictability; individual macaques could not predict where food items would precisely be thrown. This, coupled with the wide, and again unpredictable, range of size and nutritional value of particular food items, did not allow such provisioned food to be very easily monopolised. Thus, access to both, the tourists and the food items themselves, appeared to be largely indefensible under these conditions; limited contest competition could, however, occur once an individual was able to gain access to a particular item. The exact extent to which the elevated levels of feeding competition, observed during scavenging regimes, depends on the temporal and spatial availability, size and nutritional quality of the food provided, however, awaits further investigation.

12.11 Epilogue

What are likely to be the long-term effects of the provisioning of the bonnet macaques of Bandipur-Mudumalai carried out regularly by the benevolent tourists visiting the two sanctuaries and travelling beyond? What, indeed, does the future hold for them?

There have been a number of studies that have documented the detrimental effect of tourism on different primate populations due to close contact with humans; these have variously included enhanced levels of disease outbreak (reviewed in Wallis and Lee 1999), infant mortality (Berman et al. 2007), generation of stress (Kinnaird and O'Brien 1996; Butynski 2001), over-habituation and hyper-aggression (Zhao and Deng 1992; Boinski and Sirot 1997; Grossberg et al. 2003) or changes in behavioural patterns (Johns 1996; Kinnaird and O'Brien 1996; de la Torre et al. 1999; Koganezawa and Imaki 1999; O'Leary and Fa 1993; Treves and Brandon 2005) that could, in turn, affect diet, predation, inter-group relationships or even social development. A possible inkling of the difficult nature of the road ahead for our study population of bonnet macaques comes from our observation that 7 of the 21 troops that we had originally begun to study in the year 2000 have, over the last 10 years, either died out or disappeared from our study area. It is striking that these troops represent approximately 64% of the 11 unimale troops that marked the beginning of our study. In stark contrast, none of the ten original multimale troops have met with such a fate during these same years.

Our observations reveal that two sets of processes, one stochastic and the other biological, may be responsible for this differential survival of the two social organisations. First, given the small size of a typical unimale troop, the death of even a single individual, either due to predation by leopards or the occasional traffic accident, can statistically jeopardise the existence of the troop. We have thus documented at least two instances where, following the death of an adult female or the single adult male, the last remaining members of the troop have emigrated, either singly or in a group, to join other troops within range or to permanently leave the study area.

The biological processes that have also led to the dissipation of unimale troops are an important consequence of the behavioural interactions that typically characterise this form of social organisation in Bandipur-Mudumalai. The despotic behaviour of the resident male in these troops seems to promote increased male emigration from such groups and could result in the observed characteristic depletion of subadult and, occasionally, juvenile males from unimale troops (described in Sinha et al. 2005). Male emigration at a relatively young age and the newly observed emigration by females from unimale troops, driven possibly by a lack of mate choice, entail potentially heavy costs for the individual, most notably once again in the form of predation by leopards (Mukhopadhyay and Sinha, unpublished observations), and hence, for their troops. Finally, the extensive emigration of individuals of different age-sex categories from unimale troops, coupled with the regular formation of such troops by group fission, could, in turn, negatively affect the survival probability of a large section of the population, and this, in our opinion, is a matter of great conservation concern.

It is thus imperative that more detailed ecological and behavioural studies of wild, semi-urban and urban bonnet macaques are initiated urgently in order to explore other consequences of our activities on their lives and their survival. Successful management strategies also need to be developed for the species, not only in protected forest areas where, as this chapter amply demonstrates, the monkeys tend to interact increasingly with tourists to their detriment, but also in rural and urban areas where rapidly growing human populations are coming into serious conflict with the macaques with very little or almost no hope for the survival of the primates (Singh et al. 2011). We must ensure that the common bonnet macaque of today does not indeed become an endangered species of tomorrow.

Acknowledgements We owe a deep sense of gratitude to our students and colleagues, particularly Sunita Ram, Anirban Datta-Roy, Mayukh Chatterjee, Kishor Bhat, Suri Venkatachalam, Niranjan V. Joshi and Sindhu Radhakrishna, who have variously collected field data, written computer software, helped in data analysis or the writing of our papers but, above all, have shared our love for bonnet macaques and taught us much about them. We thank the Forest Departments of Karnataka and Tamil Nadu for the permission to conduct our research in the Bandipur National Park and Mudumalai Wildlife Sanctuary, respectively, and the Nature Conservation Foundation in Mysore, particularly Naveen Raj, D. H. Tanuja, Vena Kapoor, Rakhee Karumbaya, Pavithra Sankaran, M. D. Madhusudan, Aparajita Datta, Charudutt Mishra and Rohan Arthur for the valuable logistic support. Research grants from the Wenner-Gren Foundation for Anthropological Research, New York (Gr 6416 and Gr 7186), and the Department of Science and Technology, Government of India, New Delhi, are gratefully acknowledged. We thank Mewa Singh and Sindhu Radhakrishna for their insightful comments on an earlier version of this manuscript. Finally, we would like to express our appreciation for the wonderful discussions and debates, often late into the night, we have had with our friends and colleagues from the National Institute of Advanced Studies, Indian Institute of Science and the National Centre for Biological Sciences, all in Bangalore and the Nature Conservation Foundation in Mysore, without which many of our ideas would have never taken concrete shape.

References

Ali R (1981) The ecology and social behaviour of the Agastyamalai bonnet macaque. PhD thesis, Bristol University, Bristol, UK

Ali R (1984) Sexual behaviour in the southern bonnet macaque, *Macaca radiata diluta*. Int J Primatol 5:319

Altmann J, Muruthi P (1988) Differences in daily life between semi-provisioned and wild-feeding baboons. Am J Primatol 15:213–221

Barton RA (1993) Sociospatial mechanisms of feeding competition in female olive baboons, *Papio anubis*. Anim Behav 46:791–802

Berman CM, Li J, Ogawa H, Ionica C, Yin H (2007) Primate tourism, range restriction, and infant risk among *Macaca thibetana* at Mt. Huangshan, China. Int J Primatol 28:1123–1141

Boinski S, Sirot L (1997) Uncertain conservation status of squirrel monkeys in Costa Rica, *Saimiri oestdi oerstedi* and *Saimiri oerstedi citrinellus*. Folia Primatol 68:181–193

Butynski TM (2001) Africa's great apes. In: Beck B, Stoinski S, Hutchins M, Maple T, Norton B, Rowan A, Stevens E, Arluke A (eds) Great apes and humans: the ethics of coexistence. Smithsonian Institution Press, Washington, DC, pp 3–56

Chatterjee M, Mukhopadhyay K, Sinha A (2009) A many-splendored world! Long-term studies on bonnet macaques in the Bandipur National Park – Mudumalai Wildlife Sanctuary, southern India. Eighth Göttinger Freilandtage: Long-term Field Studies of Primates. German Primate Centre, Göttingen

Chauhan A, Pirta RS (2010) Socio-ecology of two species of non-human primates, rhesus monkey (*Macaca mulatta*) and Hanuman langur (*Semnopithecus entellus*), in Shimla, Himachal Pradesh. J Human Ecol 30:171–177

D'Souza L, Singh M (1992) Density and demography in roadside bonnet monkeys *Macaca radiata* around Mysore. J Ecobiol 4:87–93

de la Torre S, Snowdon CT, Bejarano M (1999) Preliminary study of the effects of ecotourism and human traffic on the howling behavior of red howler monkeys, *Alouatta seniculus*, in Ecuadorian Amazonia. Neotrop Primates 7:84–86

de Waal FBM (1989) The myth of a simple relation between space and aggression in captive primates. Zoo Biol Suppl S1:141–148

Fuentes A (2006) Human culture and monkey behavior: assessing the contexts of potential pathogen transmission between macaques and humans. Am J Primatol 68:880–896

Fuentes A, Gamerl S (2005) Disproportionate participation by age/sex classes in aggressive interactions between long-tailed macaques (*Macaca fascicularis*) and human tourists at Padangtegal Macaque Forest, Bali, Indonesia. Am J Primatol 66:197–204

Gore MA (1993) Effects of food distribution on foraging competition in rhesus monkeys, *Macaca mulatta*, and hamadryas baboons, *Papio hamadryas*. Anim Behav 45:773–786

Grossberg R, Treves T, Naughton-Treves L (2003) The incidental ecotourist: measuring visitor impacts on endangered howler monkeys at a Belizean archaeological site. Environ Conserv 30:40–51

Hill DA (1999) Effects of provisioning on the social behavior of Japanese and rhesus macaques: implications for socioecology. Primates 40:187–198

Hsu M, Kao CC, Agoramoorthy G (2009) Interactions between visitors and Formosan macaques (*Macaca cyclopis*) at Shou-Shan Nature Park, Taiwan. Am J Primatol 71:214–222

Imanishi K (1957) Social behavior in the Japanese macaque, *Macaca fuscata*. Psychologia 1:47–54

Isbell LA (1991) Contest and scramble competition: patterns of female aggression and ranging behavior among primates. Behav Ecol 2:143–155

Janson CH, van Schaik CP (1988) Recognizing the many faces of primate food competition: methods. Behaviour 105:165–186

Johns BG (1996) Responses of chimpanzees to habituation and tourism in the Kibale forest, Uganda. Biol Conserv 78:257–262

Kinnaird MF, O'Brien TG (1996) Ecotourism in the Tangkoko DuaSudara nature reserve: opening Pandora's box? Oryx 30:65–73

Koganezawa M, Imaki H (1999) The effects of food sources on Japanese monkey home range size and location, and population dynamics. Primates 40:177–185

Koyama N (1973) Dominance, grooming, and clasped-sleeping relationships among bonnet monkeys in India. Primates 14:225–244

Krishnan M (1972) An ecological survey of the larger mammals of peninsular India: Part I. J Bombay Nat Hist Soc 68:503–555

Kumar A, Kurup GU (1985) Sexual behaviour in the lion-tailed macaque, *Macaca silenus*. In: Heltne PG (ed) The lion-tailed macaques: status and conservation, vol 7, Monographs in primatology. Alan R. Liss, New York, pp 109–130

Kurup GU (1981) Report on the census surveys of rural and urban populations of non-human primates of south India. Man and biosphere programme: project no. 124. Zoological Survey of India, Calicut (now Kozhikode)

McCarthy MS, Matheson MD, Lester JD, Sheeran LK, Li J, Wagner SR (2009) Sequences of Tibetan macaque (*Macaca thibetana*) and tourist behaviors at Mt. Huangshan, China. Primate Conserv 24:145–151

Mori A (1977) Intra-troop spacing mechanism of the wild Japanese monkeys of the Koshima troop. Primates 18:331–357

Nunn CL (1999) The number of males in primate social groups: a comparative test of the socio-ecological model. Behav Ecol Sociobiol 46:1–13

O'Leary H, Fa JE (1993) Effects of tourists on Barbary macaques at Gibraltar. Folia Primatol 61:77–91

Oi T (1996) Sexual behaviour and mating system of the wild pig-tailed macaque in West Sumatra. In: Fa JE, Lindburg DG (eds) Evolution and ecology of macaque societies. Cambridge University Press, Cambridge, pp 342–368

Ram S, Venkatachalam S, Sinha A (2003) Changing social strategies of wild female bonnet macaques during natural foraging and on provisioning. Curr Sci 84:780–790

Ramanujan AK (1985) Poems of love and war: from the eight anthologies and the ten long poems of classical Tamil. Columbia University Press, New York

Ridley M (1986) The number of males in a primate troop. Anim Behav 34:1848–1858

Riley EP (2007) Flexibility in diet and activity patterns of *Macaca tonkeana* in response to anthropogenic habitat alteration. Int J Primatol 28:107–133

Riley EP (2008) Ranging patterns and habitat use of Sulawesi Tonkean macaques (*Macaca tonkeana*) in a human-modified habitat. Am J Primatol 70:670–679

Roonwal ML, Mohnot SM (1977) Primates of south Asia: ecology, socio-biology and behavior. Harvard University Press, Cambridge

Saj T, Sicotte P, Paterson JD (1999) Influence of human food consumption on the time budget of vervets. Int J Primatol 20:977–994

Schlotterhausen L (1992) Interactions between people and bonnet monkeys Macaca radiata at Alagarkoil, a south Indian temple. MA thesis, New York University, New York

Schlotterhausen L (1999) Town monkeys, country monkeys: a socioecological comparison of a human commensal and wild group of bonnet macaques (Macaca radiata). PhD thesis, New York University, New York

Simonds PE (1965) The bonnet macaque in south India. In: DeVore I (ed) Primate behavior: field studies of monkeys and apes. Holt, Winehart and Winston, New York, pp 175–196

Simonds PE (1974) Sex differences in bonnet macaque networks and social structure. Arch Sex Behav 3:151–166

Singh M, Rao NR (2004) Population dynamics and conservation of commensal bonnet macaques. Int J Primatol 25:847–859

Singh M, Akram N, Pirta RS (1984) Evolution of demographic patterns in the bonnet monkey *Macaca radiata*. In: Roonwal ML, Mohnot SM, Rathore NS (eds) Current primate researches. Zoology Department, Jodhpur University, Jodhpur, pp 7–16

Singh M, Kumara HN, Kumar MA, Singh M, Cooper M (2006) Male influx, infanticide, and female transfer in *Macaca radiata radiata*. Int J Primatol 27:515–528

Singh M, Erinjery JJ, Kavana TS, Roy K, Singh M (2011) Drastic population decline and conservation prospects of roadside dark-bellied bonnet macaques (*Macaca radiata radiata*) of southern India. Primates 52:149–154

Sinha A (2001a) The bonnet macaque revisited: ecology, demography and behaviour. ENVIS Bull Wildl Prot Areas 1:30–39

Sinha A (2001b) The monkey in the town's commons: a natural history of the Indian bonnet macaque (NIAS Report R 2–01). National Institute of Advanced Studies, Bangalore

Sinha A, Mukhopadhyay K, Datta-Roy A, Ram S (2005) Ecology proposes, behaviour disposes: ecological variability in social organisation and male behavioural strategies among wild bonnet macaques. Curr Sci 89:1166–1179

Smuts B, Cheney DL, Seyfarth RM, Wrangham RW, Struhsaker TT (eds) (1987) Primate societies. University of Chicago Press, Chicago

Southwick CH, Siddiqi FM, Farooqui YM, Pal BC (1976) Effects of artificial feeding on aggressive behaviour of rhesus monkeys in India. Anim Behav 24:11–15

Sugiyama Y (1971) Characteristics of the social life of bonnet macaques *Macaca radiata*. Primates 12:247–266

Treves A, Brandon K (2005) Tourist impacts on the behavior of black howling monkeys (Alouatta pigra) at Lamanai, Belize. In: Paterson J, Wallis J (eds) Commensalism and conflict: the human-primate interface. American Society of Primatologists, Norman

van Noordwijk MA, van Schaik CP (1987) Competition among female long-tailed macaques, *Macaca fascicularis*. Anim Behav 35:577–589

van Schaik CP (1989) The ecology of social relationships amongst female primates. In: Standen V, Foley RA (eds) Comparative socioecology: the behavioural ecology of humans and other mammals. Blackwell, Oxford, pp 195–218

Wallis J, Lee DR (1999) Primate conservation: the prevention of disease transmission. Int J Primatol 20:803–826

Watanabe B, Urasopon N, Malaivijitnond S (2007) Long-tailed macaques use human hair as dental floss. Am J Primatol 69:940–944

Wheatley BP, Harya Putra DK (1994) Biting the hand that feeds you: monkeys and tourists in Balinese monkey forests. Trop Biodivers 2:317–327

Wrangham RW (1980) An ecological model of female-bonded primate groups. Behaviour 75:262–300

Zhao QK, Deng ZY (1992) Dramatic consequences of food handouts to *Macaca thibetana* at Mount Emei, China. Folia Primatol 58:24–31

Chapter 13
Anthropogenic Influences on Macaque Populations and Their Genetic Consequences

Debapriyo Chakraborty and David Glenn Smith

13.1 Introduction

The impact of human population growth on biodiversity accelerated with the invention and spread of agriculture over the past 10,000 years (Redman 1999; McKee 2003). This lifestyle shift, from nomadic foraging by small bands of people to a sedentary lifestyle, was made possible by the development of primary food production processes, largely mono-cropping and herding techniques (Barrett et al. 1998). Predictable food supplies altered the birth-death rate equilibrium of these early human populations, resulting in increased population densities (Roberts and Manchester 1997). Although the existence of populations of other species continued to be impacted by our growing ecological influences, it was now mediated in a different way. Rather than directly killing off individuals or populations of other species, mainly through hunting or outcompeting them for natural food resources, the new agriculturists promoted wholesale displacement of both plants and animals by utilising swaths of land for crops and herding. Agricultural lands necessarily became less diverse and less productive in biomass as concentrations of domesticates were grown specifically for human consumption, at the expense of more diverse systems that had evolved to sustain different wildlife species. Although reliance on fewer food types decreased nutritional value intake, human populations managed to

D. Chakraborty (✉)
Primate Programme, Nature Conservation Foundation,
Mysore 570002, India

National Centre for Biological Sciences,
University of Agricultural Sciences Campus, Bellary Road,
Bangalore 560065, India
e-mail: debapriyo@ncbs.res.in

D.G. Smith
Department of Anthropology and California National Primate Research Center,
University of California, Davis, CA 95616, USA

S. Radhakrishna et al. (eds.), *The Macaque Connection: Cooperation and Conflict between Humans and Macaques*, Developments in Primatology: Progress and Prospects 43,
DOI 10.1007/978-1-4614-3967-7_13,

flourish at unprecedented scales (McKee 2003). Meanwhile, large mammal extinctions reached an all-time high during this time. For example, in South Africa alone, 16 species of large mammals went extinct in the past 10,000 years, including nine in historic times (McKee 1995). This is in contrast to the general pattern, prior to the emergence of the genus *Homo*, of an extinction rate of about four large mammal species every 100,000 years (McKee 1995).

Primates, the group that incorporates humans, have traditionally been considered to be the most evolutionarily successful taxon, based on several parameters including maximum lifespan potential (MLP), MLP calorie consumption (MCC), encephalisation quotient (EQ), and extra number of cortical neurons (N_c), as compared to other mammalian orders (Cutler 1976). Interestingly, however, few primate taxa occur outside the tropics, and most taxa are rare due to their small geographical range size or latitudinal extent, low population density, or both. Harcourt (2006) found that rare primate taxa are specialised, but neither use more resources nor breed more slowly than do the common ones. Instead, the correlation of rarity and specialisation was found to be mediated through geographical range: Taxa with small ranges, or small ranges for their density, are specialised, but not so taxa at low density. Furthermore, common taxa are generalised but only so because they consist of more differently specialised sub-taxa, not because each sub-taxon is generalised. One such 'generalist' and, in turn, successful non-human primate group is the genus *Macaca*. Among extant primates, the macaques, with 22 well-accepted species (Thierry 2007), occupy a geographical range that is only smaller than that of humans. In terms of their wide distribution, numerous populations and the range of habitat types exploited, macaques have achieved outstanding evolutionary success among primate groups. Although their distribution and numbers have reduced drastically since the Pleistocene, when macaques reached the peak of their evolution, the ecological adaptability and behavioural flexibility of the genus has undoubtedly contributed to their colonising success and the ability of some species to thrive in habitats undergoing drastic modification by humans (Sinha et al. 2005; Lane 2011).

Although evolutionarily successful, macaques have also been affected by anthropological influences and mostly adversely so. Undocumented historical influences of humans on the range of macaque species in the past might be reflected in the disjunction of some species distributions, origin of hybrid species or the expansion of the range of some species through introductions. Most of the well-documented cases are, however, more recent phenomena. A major part of Asia has undergone an unprecedented industrial growth and a spurt in human population size over the last 50 years. Large parts of erstwhile macaque habitats have been, as a result, converted into human habitations making some of the specialised species such as the lion-tailed macaque (*M. silenus*) of India severely endangered. Even the demand of the biomedical research community for non-human primate models, especially rhesus macaques *M. mulatta* and long-tailed macaques *M. fascicularis*, has fostered the depletion of macaques in many areas of their erstwhile distribution and their wide-scale illegal translocation from areas where captive breeding for sale is not permitted. In the present chapter, we will discuss the effect of such adverse anthropological influences on the distribution, demography, and hybridisation of macaque species from a genetic perspective.

13.2 Anthropogenic Effects on Macaque Demographic Histories

Studies in population genetics usually involve the analyses of genetic variability within and among populations and the various forces that shape it. Recent demographic events such as population bottlenecks or expansions create typical signatures of genetic diversity at selectively neutral loci, such as microsatellites or single nucleotide polymorphisms (SNPs); these are usually characterised by parameters such as the population mutation parameter, estimated, in turn, through other variables such as effective population size and rate of mutation. The comparison of different estimates of these parameters across populations, based on neutral genetic markers such as those mentioned above, can then be used to infer demographic historical events experienced by those populations.

Some studies have used this approach to infer the effect of significant environmental events, which may have occurred thousands of years ago, such as glaciations (Modolo et al. 2005), drop or rise in sea levels (Evans et al. 2003; Ziegler et al. 2007; Li et al. 2011) or volcanic activity (Hayaishi and Kawamoto 2006) on the genetic diversity and structure of macaque populations. Glaciations and subsequent warming periods, for example, have been shown to have strong effects on the phylogeography and demography of populations. During glacial advances, widely distributed populations typically shrink in size and either become extinct or retreat to fragmented pockets of where some environmental conditions favourable for life continue to prevail. There is often a severe restriction in gene flow among such pockets, leading to genetic differentiation among individuals in these different refugia (Fa and Lindburg 1996; Abegg and Thierry 2002). The resulting populations usually decline while genetic bottlenecks leave characteristic signatures in the patterns of genetic diversity among these multiple isolated populations. Relatively rare alleles are eliminated from gene pools, yielding survivor populations with an excess of intermediate frequency alleles and increased levels of genetic subdivisions, often leading to a deficiency of genetic heterozygosity within each refugium. Modolo et al. (2005), for example, reported evidence from mitochondrial DNA that suggested that the ancient ancestral population of Barbary macaques was first isolated during the onset of Pleistocene (Riss) glaciation some 213,000 years ago. The isolation deepened during the actual glaciation period and created several isolated refugia such as those currently extant in Morocco and Kherrata (Algeria), as illustrated in Fig. 13.1. A similar historical demographic pattern has been established for rhesus and long-tailed macaques (Melnick et al. 1993; Smith and McDonough 2005; Smith et al. 2007), for Japanese macaques (Hayaishi and Kawamoto 2006; Kawamoto et al. 2008a) and for Sulawesi macaques (Riley 2010).

The impact of the rise of human populations on macaques is, of course, of much more recent origin as compared to climatic events such as glaciation. Modern humans may have been present in Arabia and southern Asia earlier than previously believed and probably coincident with the documented presence of humans in the Levant sometime between 130,000 and 70,000 years ago (Petraglia et al. 2010). The kind and intensity of their interaction with macaques at that time, however, remains

Fig. 13.1 Map showing seven isolated relict populations of Barbary macaques in Algeria (*shaded*). There are other regions where macaque populations were reported a few decades ago but have since disappeared, most likely due to anthropological interference (*stippled*) (Modified from von Segesser et al. 1999)

undocumented. Reports of human–macaque interactions are known from at least the last 5,000 years (as, e.g. the classical records from the Neolithic Jomon Era of Japan, Mito and Watanabe 1999) but provide little insight into the response of the affected macaque populations to increasing human dominance across their distribution range. Furthermore, evidence of an association between the signature of a demographic event and its cause is mostly circumstantial. To confidently conclude, therefore, that a certain human activity may have caused a historical event such as a genetic bottleneck or a population expansion requires corroboration from the dating of the event using a molecular clock that could potentially display a linear correlation between levels of genetic diversity and time. In addition, such a conclusion would be justified if evidence of human intervention at the same time and place could be identified independently. The absence of early written records and the technical difficulty of differentiating the results of human intervention from environmental causes can, however, complicate such attempts. It is remarkable, nevertheless, that a few studies have indeed been able to draw confident inferences regarding the effect of human interference on macaque demographic histories (Modolo et al. 2005, 2008; Kawamoto et al. 2008a, b).

13.2.1 Effect of Habitat Destruction: Barbary Macaque Populations in Northern Africa

The Barbary macaque (*M. sylvanus*), traditionally considered the most primitive of all macaques, is the only primate left in Africa north of the Sahara desert. Once widely distributed across Europe and northern Africa, this species is now restricted to a few relict cedar forests of Algeria and Morocco (Delson 1980) where most populations occur at very low densities in heavily disturbed and fragmented habitats (Camperio Ciani et al. 1999). The important question here is whether such large-scale temporal changes in the demography of the species have left genetic signatures in its genome.

While Barbary macaque populations may have been interconnected through extended forest patches at the beginning of the Holocene, various civilisations appear to have contributed to the almost complete depletion of the once-vast lowland forests of northern Africa within a span of a 1,000 years. Fa (1984) has noted in detail how these forests have provided fuel, timber, pasture and arable land for people, resulting in almost 86% loss of forested lands in Morocco and Algeria since the time of the Romans. Such reduction in forest cover continued with the subsequent influx of the Bedouin people from the east (643–698 AD). In all lands under Arab domination, pastoralists were substituted by sedentary agriculturists, who, in effect, exerted severe pressure on the natural vegetation of the region, especially the lowland forests, but had little influence on the mountains. In these highland areas, therefore, the forests remained largely untouched since historical times. Macaque populations living in high-altitude habitats thus appear to have been less disturbed than were the more threatened lowland ones (Fa 1983).

Such a significant reduction in the extent and quality of the habitat expectedly revealed a strong signature of geographic isolation in mtDNA haplotypes among the seven present-day Barbary macaque populations illustrated (shaded) in Fig. 13.1 (von Segesser et al. 1999). Moreover, additional populations, previously reported from Algeria (stippled in Fig. 13.1), appear to have disappeared over this period of time. While all the four isolated populations under study were found to be genetically distinct, genetic polymorphisms at the microsatellite loci screened were observed to be particularly high among the surviving Algerian populations of the macaque, as compared to other genetic markers used in earlier studies (von Segesser et al. 1999). It is perhaps noteworthy that the Barbary macaque differs from all macaque species in having a highly promiscuous mating system where multiple males in a group achieve paternity successfully at any given time. This system would tend to increase the effective population size and this, in turn, may have mitigated the effect of genetic isolation (von Segesser et al. 1995) as was evident from the observed hyper-variability of the microsatellite loci.

In summary, Barbary macaque populations have clearly undergone a severe population bottleneck that mirrors the historical fragmentation of their habitat. Unexpectedly, however, the extant populations of the species reveal a genetic variability that does not match the intensity of the well-recorded anthropogenic impacts that their ancestors must have faced. Such a mismatch is often a result of the complex interplay of multiple factors including anthropogenic effects, climatic events and the more inherent biological traits of the species, the untangling of which requires long-term genetic monitoring of the populations along with a clear understanding of their reproductive behaviour and its potential genetic impacts. There is, however, hardly any study that has examined these problems adequately in any nonhuman primate system. Nevertheless, von Segesser et al. (1999) have strongly advised conserving the isolated populations of the species in order to maintain the present level of its genetic variability over the long term.

13.2.2 The Case of the Japanese Macaque: Impacts of Glaciation, Volcanic Eruptions and Hunting

The major subspecies of the Japanese macaque, *Macaca fuscata fuscata*, occupies three of the four main islands of Japan (Honshu, Shikoku and Kyushu, being absent only from Hokkaido and the Ryukyu Islands) and a number of smaller islands around these three. A second subspecies, *M. f. yakui*, inhabits the island of Yakushima, the southernmost extent of the species' distribution. The endangered status and very low levels of genetic heterogeneity displayed by this subspecies (Hayaishi and Kawamoto 2006) bespeaks its isolation and extremely limited distribution. The subspecies was hunted almost to extinction by the early twentieth century but rebounded in the post-World War II era with its populations increasing steadily since 1978.

Kawamoto et al. (2008a) found a very weak signature of population bottleneck in the northernmost Shimokita island population of *M. f. fuscata* using autosomal and Y-chromosome microsatellite loci; paleontological evidence suggests that this population has existed here over the last 120,000 years (Iwamoto and Hasegawa 1972; Aimi 2002). The last glacial maxima and the following warm period, as well as more recent hunting and forest destruction, well documented from as early as 120 years ago and which continued until the 1970s, seem to have affected this macaque population to a significant extent (Nozawa et al. 1991; Kawamoto et al. 2008a). Although the low genetic variability of the study population, reported by Kawamoto et al. (2008a), could be expected from a demographically stable population (Cornuet and Luikart 1996), the failure to detect any excess of heterozygosity could be due to the disproportionate loss of low-frequency alleles following a population bottleneck. The authors argue that any signature of population decline brought about by humans would have occurred much more recently and, therefore, be stronger in intensity. The bottleneck was thus assumed to have been an ancient one, with its signature decaying over time to an almost imperceptible level today. The demographic change responsible for the current low genetic variability in the Shimokita macaques, therefore, possibly occurred during the last warming period associated with a concomitant rise in sea levels. It is perhaps significant that this study, akin to that on the Barbary macaques, was also unable to find any genetic signature of the massive hunting, which occurred within a relatively short span of 120 years, on the genetic structure of the study populations.

13.3 Anthropogenic Introduction of Alien Macaque Species

Mammalian invasions have a fairly long history, the anthropogenic introduction of mammals into new areas being recorded at least since the beginning of the Neolithic period. Many ancient introductions involved wild species commensal with humans such as the black rat *Rattus rattus*, anthropophilous species like the wood mouse

Apodemus sylvaticus and domestic species such as the Corsican mouflon *Ovis aries* (Genovesi et al. 2009). Macaques constitute, arguably, one of the worst alien invasive species of the world (Lowe et al. 2000), with documented cases of introduced alien macaque species in novel environments over at least the last 200 years. Some species have been intentionally introduced into new territories such as rhesus macaques in the keys of the Archipelagos of the Canarreos and Camagüey in Cuba in order to create artificial colonies of semi-free-ranging macaques that would generate a supply of monkeys for institutions conducting biomedical research (Borroto-Páez 2009). There are, however, other instances where the introductions were accidental, including escaped animals or the abandonment of pet or zoo or circus monkeys. A notable example of the latter consists of between 10 and 30 Taiwanese macaques (*M. cyclopis*) that escaped from a zoo in Wakayama Prefecture, Japan, in 1955; this population had grown to 270 animals by 2003 (Kawamoto et al. 2007). Such incidents often threaten the local flora and fauna (mostly invertebrates and small mammals) due to sudden predation pressure on them brought about by the introduced macaques. In many cases, alien macaque introductions have resulted in subsequent admixture with populations of the native species effectively reducing the global population of the latter and causing a dilution of its local gene pool. These introductions also mark a bottleneck event, more specifically, a founder effect, in the history of the introduced species and this may be followed by a demographic expansion of the surviving founders of this population. An important case in point is again presented by the newly established population of Taiwanese macaques in western Japan, an alien introduction event that appears to be a major conservation threat to the local Japanese macaque population (Kawamoto et al. 2007).

13.3.1 Founder Effect: Long-Tailed Macaques on the Island of Mauritius

Long-tailed macaques were introduced to the island of Mauritius in the Indian Ocean by Portuguese or Dutch sailors about 400–500 years ago. The number of founders was probably quite small (Bonhomme et al. 2008) and considered to represent a single introduction of pets released by the sailors from an Indonesian source (Sussman and Tattersall 1986). Genetic analyses have, however, implicated multiple geographic origins for the founder population including Java (Bonhomme et al. 2008), supporting an earlier study by Tosi and Coke (2007). The local people have regarded the monkeys, numbering between 15,000 and 60,000 individuals (Bonhomme et al. 2008), as a major agricultural pest for a very long time and the Dutch decision to abandon the island in 1712 was apparently partially due to this problem (Sussman and Tattersall 1986). The demographic history of this introduced macaque population thus suggested a founder effect followed by rapid population expansion. The relatively reduced genetic diversity presented by the Mauritian long-tailed macaques, relative to other regional populations of this species, as demonstrated by their mtDNA (Smith et al. 2007), microsatellite loci (Kanthaswamy et al. 2008) and MHC alleles (Wiseman et al. 2007), was also consistent with a hypothesis of an effective genetic bottleneck. The ultimate confirmatory evidence was provided by Bonhomme et al. (2008), who showed that this population had indeed experienced a genetic bottleneck followed by at least one population expansion and possibly several cycles of later expansion and decline.

13.3.2 Effect of Genetic Variability on Population Survivorship

Classically, a population bottleneck, followed by a founder effect, is thought to give rise to low genetic diversity through a small founder size and eventual inbreeding (Wright 1931; Slatkin 1993; Frankham 1996; Frankham and Ralls 1998; Frankham et al. 2002). Such loss of genetic diversity is intimately associated with an increased risk of inbreeding depression and has been shown, in some cases, to result in decreased growth rates, fertility, fecundity and offspring viability (Keller 1998; Westemeier et al. 1998; Madsen et al. 1999). Populations that have lost genetic diversity may also suffer from an increased probability of extinction as a consequence of increased vulnerability to novel pathogens. The maintenance of genetic diversity has thus been traditionally considered to be of fundamental importance in conservation biology (Frankham 1996; Frankham and Ralls 1998; Madsen et al. 1999). There are, nevertheless, reports of an increasing number of species for which the extent of genetic variability and the ability to respond to diseases or environmental change differ markedly from expectations (Amos and Harwood 1998).

An extremely important factor that influences immunocompetence or the ability of an organism to defend itself from disease consists of the major histocompatibility complex (MHC) genes possessed by that organism. MHC genes are responsible for adaptive immune responses in vertebrates and are, thereby, involved in modulating host resistance to emerging pathogens (Klein 1986). Most natural animal populations exhibit high MHC diversity in terms of the number of alleles present, the extent of sequence variation among these alleles and the levels of heterozygosity (Klein 1986; Hedrick 2003). Very high diversity at MHC loci among and within vertebrate species has been interpreted as adaptations to detect and present a wide array of peptides from rapidly evolving pathogens (Garrigan and Hedrick 2003) whose distributions are often geography-specific. Consequently, high levels of MHC diversity could be the ultimate response of natural selection to unpredictable or temporally varying disease outbreaks (Altizer et al. 2003). Some genetically bottlenecked animal populations including, for example, Scandinavian beavers (*Castor fiber*; Ellegren et al. 1993), fallow deer (*Cervus dama*; Mikko et al. 1999), northern elephant seals (*Mirounga angustirostris*; Weber et al. 2004) and African cheetahs (*Acinonyx jubatus*; Castro-Prieto et al. 2011), however, exhibit low or no detectable polymorphisms in MHC genes but yet have survived, or even increased in numbers, with no apparent increase in susceptibility to infectious diseases (but see

Radwan et al. 2010). In sheer contrast, bottlenecked populations of desert bighorn sheep (*Ovis aries*) are highly susceptible to many infectious diseases despite high levels of MHC diversity (Gutierrez-Espeleta et al. 2001). The precise implications of the extent of genetic diversity at MHC loci for the survivorship of non-primate mammalian species thus remain unclear. Let us now turn to the results of a similar study on the aforementioned introduced population of long-tailed macaques on the island of Mauritius.

The Mauritian long-tailed macaques are particularly valuable for detailed immunological studies due to their remarkably restricted MHC diversity (Wiseman et al. 2007; Mee et al. 2009). Bonhomme et al. (2008) compared the variation of microsatellite loci within and outside the MHC region in these macaques with that of natural populations from the Philippines, Java and some areas of southeast Asia. More polymorphism was expected at the MHC microsatellites as they are closely linked to the functional MHC genes than are the non-MHC microsatellites as a result of balancing selection working on the MHC genes. All the study populations, however, displayed similar levels of genetic diversity at both MHC and non-MHC microsatellites while none of the MHC microsatellites from the Mauritian individuals were found to be under positive or balancing selection. It has thus been suggested that a rapid demographic expansion possibly followed the bottleneck event created by the introduction of the founder individuals (Lawler et al. 1995; Sussman and Tattersall 1986) and this maintained the current substantial and comparable levels of genetic diversity at both the MHC and non-MHC loci without the help of any kind of selection. Bonhomme et al. (2008), however, found high frequencies of some microsatellite alleles in the long-tailed macaque population from the Philippines in agreement with a pattern of directional selection on MHC class II genes. Interestingly, unlike the Mauritian individuals, those from the Philippines are able to survive infection by the malaria parasite *Plasmodium coatneyi* (Migot-Nabias et al. 1999). *Plasmodium* species represent one source of pathogens that may play a role in such directional selection and have even been hypothesised to have influenced speciation between rhesus and long-tailed macaques (Wheatley 1980). It was speculated that some MHC alleles or haplotypes (class II DRB and probably class I), prevailing in the Philippines population, play a role in the greater resistance of the macaques of this population to various *Plasmodium* parasites (Bonhomme et al. 2007). Our current understanding thus suggests that the Mauritian population continues to be susceptible to malaria in spite of their high variability at the MHC loci possibly because these individuals have not been exposed to malaria parasites on Mauritius.

13.3.3 Interspecific Hybridisations in the Wild

Ancient hybridisation events across sympatric macaque species are well known (Zinner et al. 2010). In recent times, human intervention has brought together many macaque species that were historically separated by relatively great distances thus resulting in polyspecific associations in the form of mixed-species troops and hybrids. This has been reported for rhesus macaque × long-tailed macaque and long-tailed macaque × pig-tailed macaque *M. nemestrina* crosses in Malaysia, Tonkean macaque *M. tonkeana* × Heck's macaque *M. hecki*, moor macaque *M. maura* × Tonkean macaque, Tonkean macaque × booted macaque *M. ochreata*, Celebes crested macaque *M. nigra* × Heck's macaque and Gorontalo macaque *M. nigrescens* × Heck's macaque crosses in Sulawesi (Bernstein 1966, 1968; Bynum 2002; Fooden 2006; Supriatna et al. 1992; Watanabe et al. 1991a, b; Watanabe and Matsumura 1991). Malaivijitnond and Hamada (2008) reported an interspecific mating between a released male pig-tailed macaque and female rhesus macaques in a small isolated semi-wild troop of rhesus macaques in northeastern Thailand. Rhesus macaque populations in Thailand are presently rare and isolated from one another due to increasing habitat fragmentation; such isolation could make these populations particularly vulnerable to unwanted hybridisation with released alien macaque species. Finally, mixed troops of rhesus and bonnet macaques *M. radiata* have also been reported along their common distribution boundary in central and southern India although it is uncertain whether any hybridisation has ever occurred between members of the two species in such groups (Kumar et al. 2011).

Theoretical studies have shown that a disturbance that weakens pre-mating isolation between incipient ecological species can sometimes lead to species collapse through hybridisation (Gilman and Behm 2011). Whether a particular species pair collapses is likely to depend on the strength, speed and duration of the disturbance, as well as on the genetic architecture underlying resource use and mate choice in the species pair. In the light of such insights but in the absence of any significant ecological and genetic information on the macaque populations mentioned here, it is difficult to assess the frequency and genetic impacts of human-driven natural hybridisation and its conservation implications for these species.

13.4 Human-Driven Range Expansion of Macaques

The geographic areas occupied by species are often highly dynamic, changing through time, with periods of directional or isotropic growth, of range expansions and of contractions followed by re-expansions (Taberlet and Cheddadi 2002). Range expansions have occurred repeatedly in the history of most, if not all, species and appear to be continuing even today, possibly at an increasing rate owing to rapid climatic changes. Species ranges can also occasionally expand and distribution patterns change due to human interventions. The consequences of range expansion of one species for other closely related species in the communities that are invaded can be considerable. Range expansions are also of interest because they often signal important changes in habitats, may bring together populations that have partially or just completed the process of speciation and may have interesting consequences for the communities that are invaded.

In a recent detailed study, Kumar et al. (2011) documented changes in the distribution of rhesus macaques *M. mulatta* in southeastern India; the species appears to have expanded its geographical range by ca. 3,500 km^2 since the last macaque distribution surveys in that part of the subcontinent (Fooden et al. 1981). Thirty years ago, rhesus macaques invaded the Vijayawada Hills in the southern Indian state of Andhra Pradesh, formerly occupied by bonnet macaques *M. radiata* (Fooden et al. 1981), extending its southern distributional limits. Today, the species appears to have advanced further southwards, having crossed the rivers Godavari and Krishna to intrude into the bonnet macaque distributional range. This distributional overlap region, extending to a maximum width of ca. 45 km, is now virtually devoid of bonnet macaque troops. The pace of irrigation development in the basin of the Krishna River accelerated significantly during the last 30 years, and several dams and bridges have been constructed across the river (Venot et al. 2008). This resulted in the development of large swathes of agricultural lands in the Krishna Basin (Guntur district of Andhra Pradesh) facilitating the expansion of rhesus macaques south of the river into the Guntur and Prakasam districts of the state (Kumar et al. 2011). In addition, the introduction of large numbers of rhesus macaque groups into the eastern and the western flanks of the bonnet macaque range during the last three decades has augmented range expansion by the former and local extinctions of the bonnet macaque populations from this region (Kumar et al. 2011). The larger rhesus macaques are more dominant to the comparatively smaller bonnet macaques and displace them by aggressively driving them away from food sources or preferred habitats (Kumar and Sinha 2011, personal communication). Such dominance by large-bodied primate species of smaller-bodied species and eventual displacement or exclusion of the latter from food resources is well known (Strier 2007) and has been clearly documented elsewhere too as, for example, during interspecific interactions of Neotropical primates including howlers, capuchins and muriquis (Dias and Strier 2000).

Although range expansions have occurred recurrently in the history of most species, their genetic consequences have been little investigated. Theoretical studies show that range expansions are quite different from pure demographic expansions and generally lead to a loss of genetic diversity along the expansion axis owing to recurrent bottleneck effects (Excoffier et al. 2009). Extensive migration between neighbouring populations, however, preserves genetic diversity and leads to consequences of range expansions that are more akin to those of large demographic expansions. Such range expansions appear to be conditioned by the properties and the dynamics of the wave front; the frequency of rare variants in the advancing surf front will determine the genetic structuring of the colonising population and certain distinct sectors of low genetic diversity could appear. There could also be a subsequent introgression of local genes into the genome of the invading species. Interestingly, theoretical analyses indicate that the patterns of genetic diversity generated by such an invading population could be similar to those generated by selective sweeps in an established population and could, thus, be mistakenly interpreted as adaptive events (Currat et al. 2006).

13.5 Conclusions

Human–macaque interactions constitute a complex phenomenon influencing perhaps the biology of the macaque more profoundly than ours. At the population level, humans tend to influence the distribution, demography, immunology and even behaviour of the macaque species they interact with though none of these interactions are ever simple. These work at different levels, interacting, in turn, with other environmental factors. For example, when a macaque population is exposed to a novel disease from an anthropogenic source, its distribution range might contract due to outbreak deaths, which will also have severe demographic consequences. At another level, however, this population could also consequently acquire new genetic variation through its survivors, thus influencing the immune system of its individuals. It is thus clear that the distribution, ecology and behaviour of human populations will have multiple effects on commensal and even wild macaque populations at different organisational levels (see Chap. 12 by Sinha and Mukhopadhyay, this volume for the consequences of anthropogenic impacts on macaque social structures), and most of these impacts are likely to have genetic consequences over the long term. Our current state of knowledge, unfortunately, suffers from a serious lack of insight into such genetic impacts. There is, therefore, a dire need for long-term genetic monitoring programmes to understand the effect of anthropogenic factors on the dispersal and demography of different macaque species. New laboratory and statistical techniques are enabling the use of molecular markers for genetic monitoring of wild populations; effective studies, however, are still very few, even on well-studied taxa (Kruckenhauser et al. 2009). One of the reasons attributed to the tenuous survival displayed by macaques against the onslaught of development is their remarkable capacity to adapt to human-modified landscapes. It seems, however, that time is running out for even many of the most adaptable members of this unique group of primates.

Acknowledgements We thank Anindya Sinha and Shreejata Gupta for their positive criticism and useful suggestions that deeply improved an earlier version of this chapter. DC is also grateful to Uma Ramakrishnan for financial support in the form of a senior research fellowship during the writing of this chapter.

References

Abegg C, Thierry B (2002) Macaque evolution and dispersal in insular southeastAsia. Biol J Linn Soc 75:555–576

Aimi M (2002) The oldest fossil macaque from Japan. Primate Res 18:239–245 (In Japanese with English summary)

Altizer S, Harvell D, Friedle E (2003) Rapid evolutionary dynamics and disease threats to biodiversity. Trends Ecol Evol 18:589–596

Amos W, Harwood J (1998) Factors affecting levels of genetic diversity in natural populations. Philos Trans R Soc Lond B 353:177–186

Barrett R, Kuzawa CW, McDade T, Armelagos GJ (1998) Emerging and re-emerging infectious diseases: the third epidemiologic transition. Annu Rev Anthropol 27:247–271
Bernstein IS (1966) Naturally occurring primate hybrid. Science 154:1559–1560
Bernstein IS (1968) Social status of two hybrids in a wild troop of *Macaca irus*. Folia Primatol 8:121–131
Bonhomme M, Blancher A, Jalil MF, Crouau-Roy B (2007) Factors shaping genetic variation in the MHC of natural non-human primate populations. Tissue Antigens 70:398–411
Bonhomme M, Blancher A, Cuartero S, Chikhi L, Crouau-Roy B (2008) Origin and number of founders in an introduced insular primate: estimation from nuclear genetic data. Mol Ecol 17:1009–1019
Borroto-Páez R (2009) Invasive mammals in Cuba: an overview. Biol Invasions 11:2279–2290
Bynum EL (2002) Morphological variation within a macaque hybrid zone. Am J Phys Anthropol 118:45–49
Camperio Ciani A, Mouna M, Arhou M (1999) *Macaca sylvanus* as a biological indicator of the cedar forest quality. In: Selected proceedings of the first international conference on biodiversity and natural resources preservation. Al Akhawayn University Press, Ifrane, pp 91–98
Castro-Prieto A, Wachter B, Sommer S (2011) Cheetah paradigm revisited: MHC diversity in the largest free-ranging population. Mol Biol Evol 28:1455–1468
Cornuet JM, Luikart G (1996) Description and power analysis of two tests for detecting recent population bottlenecks from allele frequency data. Genetics 144:2001–2014
Currat M, Excoffier L, Maddison W, Otto SP, Ray N, Whitlock MC, Yeaman S (2006) Comment on "Ongoing adaptive evolution of ASPM, a brain size determinant in *Homo sapiens*" and "Microcephalin, a gene regulating brain size, continues to evolve adaptively in humans. Science 313:172a
Cutler RG (1976) Evolution of longevity in primates. J Hum Evol 5:169–202
Delson E (1980) Fossil macaques, phyletic relationships and a scenario of deployment. In: Lindburg E (ed) The macaques: studies in ecology, behavior and evolution. Van Nostrand, New York
Dias LG, Strier KB (2000) Agonistic encounters between muriquis, *Brachyteles arachnoides hypoxanthus* (Primates, Cebidae), and other animals at the Estacão Biológica de Caratinga, Minas Gerais, Brazil. Neotrop Primates 8:138–141
Ellegren H, Hartman G, Johansson M, Andersson L (1993) Major histocompatibility complex monomorphism and low-levels of DNA-fingerprinting variability in a reintroduced and rapidly expanding population of beavers. Proc Natl Acad Sci USA 90:8150–8153
Evans BJ, Supriatna J, Andayani N, Melnick DJ (2003) Diversification of Sulawesi macaque monkeys: decoupled evolution of mitochondrial and autosomal DNA. Evolution 57:1931–1946
Excoffier L, Foll M, Petit RJ (2009) Genetic consequences of range expansions. Annu Rev Ecol Evol Syst 40:481–501
Fa JE (1983) An analysis of the status of the Barbary macaque (*Macaca sylvanus,* L.) in the wild: steps towards its conservation. In: Harper D (ed) Symposium on the conservation of primates and their habitats, Vaughan papers. University of Leicester, Leicester
Fa JE (1984) Habitat distribution and habitat preference in Barbary macaques (*Macaca sylvanus*). Int J Primatol 5:273–286
Fa JE, Lindburg DG (1996) Evolution and ecology of macaque societies. Cambridge University Press, Cambridge
Fooden J (2006) Comparative review of the *fascicularis*-group species of macaques (Primates: *Macaca*). Fieldiana Zool 107:1–43
Fooden J, Mahabal A, Saha SS (1981) Redefinition of rhesus macaque-bonnet macaque boundary in peninsular India. J Bombay Nat Hist Soc 78:463–474
Frankham R (1996) Relationship of genetic variation to population size in wildlife. Conserv Biol 10:1500–1508
Frankham R, Ralls K (1998) Inbreeding leads to extinction. Nature 92:1–2
Frankham R, Ballou JD, Briscoe DA (2002) Introduction to conservation genetics. Cambridge University Press, Cambridge

Garrigan D, Hedrick PW (2003) Detecting adaptive molecular polymorphism: lessons from the MHC. Evolution 57:1707–1722

Genovesi P, Bacher S, Kobelt M, Pascal M, Scalera R (2009) Alien mammals of Europe. Handb Alien Species Eur 3:119–128

Gilman RT, Behm JE (2011) Hybridization, species collapse, and species reemergence after disturbance to premating mechanisms of reproductive isolation. Evolution 65:2421–2724

Gutierrez-Espeleta GA, Hedrick PW, Kalinowski ST, Garrigan D, Boyce WM (2001) Is the decline of desert bighorn sheep from infectious disease the result of low MHC variation? Heredity 86:439–450

Harcourt AH (2006) Rarity in the tropics: biogeography and macroecology of the primates. J Biogeogr 33:2077–2087

Hayaishi S, Kawamoto Y (2006) Low genetic diversity and biased distribution of mitochondrial DNA haplotypes in the Japanese macaque (*Macaca fuscata yakui*) on Yakushima Island. Primates 47:158–164

Hedrick PW (2003) A heterozygote advantage. Science 302:57

Iwamoto M, Hasegawa Y (1972) Two macaque fossil teeth from the Japanese Pleistocene. Primates 13:77–81

Kanthaswamy S, Satkoski J, George D, Kou A, Erickson BJ, Smith DG (2008) Hybridization and stratification of nuclear genetic variation in *Macaca mulatta* and *Macaca fascicularis*. Int J Primatol 29:1295–1311

Kawamoto Y, Kawamoto S, Kawai S, Shirai K, Yoshida A, Hagihara K, Shiratori D, Naoi Y (2007) Status report of hybridization in an introduced population of rhesus macaques (*Macaca mulatta*) with native Japanese macaques (*M. fuscata*) in the Bousou Peninsula, Chiba, Japan. Primate Res 23:81–89 (in Japanese)

Kawamoto Y, Tomari K, Kawai S, Kawamoto S (2008a) Genetics of the Shimokita macaque population suggest an ancient bottleneck. Primates 49:32–40

Kawamoto Y, Kawamoto S, Matsubayashi K, Nozawa K, Watanabe T, Stanley MA, Perwitasari-Farajallah D (2008b) Genetic diversity of long-tailed macaques (*Macaca fascicularis*) on the island of Mauritius: an assessment of nuclear and mitochondrial DNA polymorphisms. J Med Primatol 37:45–54

Keller LF (1998) Inbreeding and its fitness effects in an insular population of song sparrows (*Melospiza melodia*). Evolution 52:240–250

Klein J (1986) Natural history of the major histocompatibility complex. Wiley, New York

Kruckenhauser L, Rauer G, Daübl B, Haring E (2009) Genetic monitoring of a founder population of brown bears (*Ursus arctos*) in central Austria. Conserv Genet 10:1223–1233

Kumar R, Radhakrishna S, Sinha A (2011) Of least concern? Range extension by rhesus macaques (*Macaca mulatta*) threatens long-term survival of bonnet macaques (*M. radiata*) in peninsular India. Int J Primatol 32:945–959

Lane KE (2011) Landscape dynamics: genetics and parasitism in Balinese long-tailed macaques (*Macaca fascicularis*). PhD dissertation, University of Notre Dame, South Bend

Lawler SH, Sussman RW, Taylor LL (1995) Mitochondrial DNA of the Mauritian macaques (*Macaca fascicularis*): an example of the founder effect. Am J Phys Anthropol 96:133–141

Li D, Xu H, Cheng A, Smith DG, Satkoski JA, Yao Y, Du D, Ni Q (2011) Phylogenetic analysis of Chinese rhesus macaques (*Macaca mulatta*) based on mitochondrial control region sequences. Am J Primatol 73:883–895

Lowe SJ, Browne M, Boudjelas S (2000) 100 of the world's worst invasive alien species. IUCN/SSC Invasive Species Specialist Group (ISSG), Auckland

Madsen T, Olsson M, Shine R, Wittzell H (1999) Restoration of an inbred adder population. Nature 402:34–35

Malaivijitnond S, Hamada Y (2008) Current situation and status of long-tailed macaques (*Macaca fascicularis*) in Thailand. Nat Hist J Chulalongkorn Univ 8:185–204

McKee JK (1995) Turnover patterns and species longevity of large mammals from the late Pliocene and Pleistocene of southern Africa: a comparison of simulated and empirical data. J Theor Biol 172:141–147

McKee JK (2003) Sparing nature: the conflict between human population growth and Earth's biodiversity. Rutgers University Press, Piscataway

Mee ET, Badhan A, Karl JA, Wiseman RW, Cutler K, Knapp LA, Almond N, O'Connor DH, Rose NJ (2009) MHC haplotype frequencies in a UK breeding colony of Mauritian cynomolgus macaques mirror those found in a distinct population from the same geographic origin. J Med Primatol 38:1–14

Melnick DJ, Hoelzer GA, Absher R, Ashley AI (1993) MtDNA diversity in rhesus monkeys reveals overestimates of divergence time and paraphyly with neighboring species. Mol Biol Evol 10:282–295

Migot-Nabias F, Ollomo B, Dubreuil G, Morelli A, Domarle O, Nabias R, Georges AJ, Millet P (1999) *Plasmodium coatneyi*: differential clinical and immune responses of two populations of *Macaca fascicularis* from different origins. Exp Parasitol 91:30–39

Mikko S, Roed K, Schmutz S, Andersson L (1999) Monomorphism and polymorphism at MHC DRB loci in domestic and wild ruminants. Immunol Pap Rev 167:169–178

Mito Y, Watanabe K (1999) The social history of people and monkeys [Hito to saru no shakai-shi]. Tokai University Press, Tokyo (In Japanese)

Modolo L, Salzburger W, Martin RD (2005) Phylogeography of Barbary macaques (*Macaca sylvanus*) and the origin of the Gibraltar colony. Proc Natl Acad Sci USA 102:7392–7397

Modolo L, Martin RD, Van Schaik CP, Van Noordwijk MA, Krützen M (2008) When dispersal fails: unexpected genetic separation in Gibraltar macaques (*Macaca sylvanus*). Mol Ecol 17:4027–4038

Nozawa K, Shotake T, Minezawa M, Kawamoto Y, Hayasaka K, Kawamoto S, Ito S (1991) Population genetics of Japanese monkeys: III. Ancestry and differentiation of local populations. Primates 32:411–435

Petraglia MD, Haslam M, Fuller DQ, Boivin N, Clarkson C (2010) Out of Africa: new hypotheses and evidence for the dispersal of *Homo sapiens* along the Indian Ocean rim. Ann Hum Biol 37:288–311

Radwan J, Biedrzycka A, Babik W (2010) Does reduced MHC diversity decrease viability of vertebrate populations? Biol Conserv 143:537–544

Redman CL (1999) Human impact on ancient environments. The University of Arizona Press, Tucson

Riley EP (2010) The endemic seven: four decades of research on the Sulawesi macaques. Evol Anthropol 19:22–36

Roberts C, Manchester K (1997) The archaeology of disease. Cornell University Press, Ithaca

Sinha A, Datta A, Madhusudan MD, Mishra C (2005) *Macaca munzala*: a new species from western Arunachal Pradesh, northeastern India. Int J Primatol 26:977–989

Slatkin M (1993) Isolation by distance in equilibrium and nonequilibrium populations. Evolution 47:264–279

Smith DG, McDonough JW (2005) Mitochondrial DNA variation in Chinese and Indian rhesus macaques (*Macaca mulatta*). Am J Primatol 65:1–25

Smith DG, Mcdonough JW, George DA (2007) Mitochondrial DNA variation within and among regional populations of longtail Macaques (*Macaca fascicularis*) in relation to other species of the *fascicularis* group of macaques. Am J Primatol 69:182–198

Strier KB (2007) Primate behavioral ecology. Allyn and Bacon, Needham Heights

Supriatna J, Froehlich JW, Erwin JM, Southwick CH (1992) Population, habitat and conservation status of *Macaca maurus*, *Macaca tonkeana* and their putative hybrids. Trop Biodivers 1:31–48

Sussman RW, Tattersall I (1986) Distribution, abundance, and putative ecological strategy of *Macaca fascicularis* on the island of Mauritius, southwestern Indian Ocean. Folia Primatol 46:28–43

Taberlet P, Cheddadi R (2002) Quaternary refugia and persistence of biodiversity. Science 297:2009–2010

Thierry B (2007) Unity in diversity: lessons from macaque societies. Evol Anthropol 16:224–238

Tosi AJ, Coke CS (2007) Comparative phylogenetics offers new insights into the biogeographic history of *Macaca fascicularis* and the origin of the Mauritian macaques. Mol Phylogenet Evol 42:498–504

Venot JP, Sharma BR, Rao KVGK (2008) The Lower Krishna Basin trajectory: relationships between basin development and downstream environmental degradation. IWMI research report 125. International Water Management Institute, Colombo

von Segesser F, Scheffrahn W, Martin RD (1995) Parentage analysis within a semi-free-ranging group of Barbary macaques *Macaca sylvanus*. Mol Ecol 4:115–120

von Segesser F, Menard N, Gaci B, Martin RD (1999) Genetic differentiation within and between isolated Algerian subpopulations of Barbary macaques (*Macaca sylvanus*): evidence from microsatellites. Mol Ecol 8:433–442

Watanabe K, Matsumura S (1991) The borderland of possible hybrids between three species of macaques, *M. nigra*, *M. nigrescens*, and *M. hecki*, in the northern peninsula of Sulawesi. Primates 32:365–369

Watanabe K, Lapasere H, Tantu R (1991a) External characteristics and associated developmental changes in two species of Sulawesi macaques, *Macaca tonkeana* and *M. hecki*, with special reference to hybrids and the borderland between the species. Primates 32:61–76

Watanabe K, Matsumura S, Watanabe T, Hamada Y (1991b) Distribution and possible intergradations between *Macaca tonkeana* and *M. ochreata* in the borderland of the species in Sulawesi. Primates 32:385–389

Weber DS, Stewart BS, Schienman J, Lehman N (2004) Major histocompatibility complex variation at three class II loci in the northern elephant seal. Mol Ecol 13:711–718

Westemeier RL, Brawn JD, Simpson SA, Esker TL, Jansen RW, Walk JW, Kershner EL, Bouzat JL, Paige KN (1998) Tracking the long-term decline and recovery of an isolated population. Science 282:1695–1698

Wheatley BP (1980) Malaria as a possible selective factor in the speciation of macaques. J Mammal 61:307–311

Wiseman RW, Wojcechowskyj JA, Greene JM, Blasky AJ, Gopon T, Soma T, Friedrich TC, O'Connor SL, O'Connor DH (2007) Simian immunodeficiency virus SIVmac239 infection of major histocompatibility complex-identical cynomolgus macaques from Mauritius. J Virol 81:349–361

Wright S (1931) Evolution in Mendelian populations. Genetics 16:97–159

Ziegler T, Abegg C, Meijaard E, Perwitasari-Farajallah D, Walter L, Keith Hodges J, Roos C (2007) Molecular phylogeny and evolutionary history of southeast Asian macaques forming the *M. silenus* group. Mol Phylogenet Evol 42:807–816

Zinner D, Arnold ML, Roos C (2010) The strange blood: natural hybridization in primates. Evol Anthropol 20:96–103

Chapter 14
Managing Humans, Managing Macaques: Human–Macaque Conflict in Asia and Africa

Nancy E.C. Priston and Matthew R. McLennan

14.1 Introduction

Conflict between humans and wild animals is one of the greatest challenges to biodiversity conservation globally (Hill et al. 2002; Woodroffe et al. 2005). Expanding human populations and large-scale, accelerating conversion of natural habitats to alternative land uses mean that wildlife populations must adapt to human-dominated environments or disappear. At the same time, sustainable coexistence between people and wildlife in shared landscapes demands that humans share space and resources with wild animals. Today, the majority of non-human primates (hereafter 'primates') are severely threatened by habitat loss and modification (Chapman and Peres 2001; Cowlishaw and Dunbar 2000). However, the responses of individual taxa to increasingly 'agriculturalised' – and, in some circumstances, urbanised – environments occur along a gradient ranging from local extinction (inability to adapt) to apparent benefit (ecological and behavioural adaptation) (Gautier and Biquand 1994). Taken as a whole, the genus *Macaca* appears to typify this latter response, though considerable variation exists among macaques in their propensity to exploit anthropogenic environments and coexist with people. Due to this frequent association with humans, macaques feature prominently in the growing database of primate–people conflicts. In this chapter, we review the human–macaque conflict situation in Asia and Africa. Using specific case studies, we explore the influence of

N.E.C. Priston (✉) • M.R. McLennan
Faculty of Humanities and Social Sciences, Department of Social Sciences,
Oxford Brookes University, Oxford OX3 0BP, UK
e-mail: nancy@cantab.net

S. Radhakrishna et al. (eds.), *The Macaque Connection: Cooperation and Conflict between Humans and Macaques*, Developments in Primatology: Progress and Prospects 43,
DOI 10.1007/978-1-4614-3967-7_14, © Springer Science+Business Media, LLC 2013

cultural factors and differing interaction contexts on conflict scenarios and the influence of policy on mitigation strategies and conservation.

Macaques are among the most successful primate radiations (Fooden 1980). Aside from humans, they occupy the widest geographical range of any primate genus: natural populations occur in North Africa, throughout south and southeast Asia and southern China, and northeast to Japan, where *M. fuscata* has the most northerly and easterly range of any primate. Although formerly widespread in Europe, today the genus is represented by a small introduced population of Barbary macaques on Gibraltar (Abegg 2006). Macaques have also been introduced to regions far outside their natural range, for example, *M. mulatta* in Puerto Rico (Engeman et al. 2010) and *M. fascicularis* in Mauritius (Sussman et al. 2011) and in the Republic of Palau (Wheatley 2011). With a latitudinal range spanning 20°S to 40°N, macaques occupy a diversity of habitats, including grasslands, mangroves, tropical rainforests, deciduous and temperate forests, rocky cliffs and coastal regions, as well as anthropogenic habitats (Thierry 2011). Given this remarkable ecological adaptability, it is unsurprising that macaques are among the most taxonomically diverse of extant primate genera, with up to 23 species in five species groups currently recognised (Fooden 1980; Groves 2001; IUCN 2011; Ziegler et al. 2007; Table 14.1).

14.1.1 Macaques as 'Pest' Primates

The last two decades have seen a proliferation of studies reporting conflict between humans and wild primates, particularly in the context of crop raiding (Hill 2005). Crops offer energetic advantages over many natural foods for primates in human-modified habitats (Forthman-Quick and Demment 1988). As natural habitats are converted to farmland, the list of primates decried as crop pests continues to grow (review in Lee and Priston 2005). Species now known to supplement their natural diets with human foods include Asian colobines (e.g. *Presbytis thomasi*, Marchal and Hill 2009), Neotropical cebids (e.g. *Cebus libidinosus*, de Freitas et al. 2008), Madagascan lemurs (*Lemur catta*, LaFleur and Gould 2009) and African great apes (e.g. *Pan troglodytes*, McLennan 2008). However, few such taxa can be considered to prosper in disturbed habitats in association with people (Richard et al. 1989). Three groups of primate are particularly successful at exploiting the human–primate interface. These are the baboons (*Papio* spp.), vervet and tantalus monkeys (*Chlorocebus* spp.) in Africa and the macaques in Asia. (Wild Barbary macaques *M. sylvanus* in North Africa appear to fare less well in anthropogenic habitats compared to many members of their genus in Asia; Mouna and Camperio Ciani 2006) Broadly speaking, members of these genera exhibit a range of traits that enable them to exploit agricultural landscapes, notably, semi-terrestrial locomotion, large and complex social groupings, flexible and varied diets, intelligence, manual dexterity and agility, and a somewhat feisty and audacious temperament (Else 1991; Knight 1999; Strum 1994).

Table 14.1 Overview of macaque–human conflict situations and conservation status (in 2000 and 2008) in 23 currently recognised species of *Macaca* in Asia and Africa

		IUCN status[1]			
Species	Distribution	2000	2008	Conflict situation	References
Barbary macaque *M. sylvanus*	Algeria, Morocco (introduced – Gibraltar)	V	E	Tourism/provisioning	Fa (1992), Fuentes (2006b), and Maréchal et al. (2011)
				Urban nuisance	Cortes and Shaw (2006) and Deag (1977)
				Crop-raiding	Camperio Ciani et al. (2001)
				Bark-stripping	Mouna and Camperio Ciani (2006)
Lion-tailed macaque *M. silenus*	India (Western Ghats)	E	E	Crop-raiding	Menon and Poirier (1996)
Crested macaque *M. nigra*	Sulawesi	E	CE	Crop-raiding	Bynum et al. (1999)
Gorontalo macaque *M. nigrescens*	Sulawesi	NT	V	Crop-raiding	Bynum et al. (1999)
Heck's macaque *M. hecki*	Sulawesi	NT	V	Crop-raiding	Bynum et al. (1999)
Tonkean macaque *M. tonkeana*	Sulawesi	NT	V	Crop-raiding	Supriatna et al. (1992), Riley (2007), and Riley and Priston (2010)
Moor macaque *M. maura*	Sulawesi	E	E	Crop-raiding	Supriatna et al. (1992)
Booted macaque *M. ochreata*	Sulawesi	DD	V	Crop-raiding	Bynum et al. (1999) and Riley et al. (2007)
Buton macaque *M. brunnescens*	Sulawesi	V	V	Crop-raiding	Priston (2005), Priston (2009), and Riley and Priston (2010)
Pagai macaque *M. pagensis*	Mentawai Islands	CE	CE	Crop-raiding	Fuentes (2002)

(continued)

Table 14.1 (continued)

Species	Distribution	IUCN status[1] 2000	IUCN status[1] 2008	Conflict situation	References
Siberut macaque *M. siberu*	Mentawai Islands	CE	V	Crop-raiding	Fuentes (2002)
Southern pig-tailed macaque *M. nemestrina*	Indonesia, Malaysia, Brunei, Thailand	V	V	Crop-raiding	Hashim et al. (2009) and Linkie et al. (2007)
Northern pig-tailed macaque *M. leonina*	Bangladesh, Cambodia, China, India, Lao PDR, Myanmar, Thailand, Vietnam	V	V	Crop-raiding	Choudhury (2003) and Srivastava (2006)
Toque macaque *M. sinica*	Sri Lanka	V	E	Crop-raiding	Nahallage et al. (2008) and Nijman and Nekaris (2010)
Bonnet macaque *M. radiata*	India	LC	LC	Crop-raiding	Chakravarthy and Thyagaraj (2005), Singh and Rao (2004), and Singh et al. (2011)
Assamese macaque *M. assamensis*	Bangladesh, Bhutan, China; India, Lao PDR, Myanmar, Nepal, Thailand, Vietnam	V	NT	Crop-raiding	Aggimarangsee (1992), Chalise and Johnson (2005), and Medhi et al. (2007)
Arunachal macaque *M. munzala*	India (Arunachal Pradesh)	–	E	Crop-raiding	Sinha et al. (2006)
Tibetan macaque *M. thibetana*	China	LC	NT	Tourism/provisioning	McCarthy et al. (2009), Ruesto et al. (2010), and Zhao and Deng (1992)
				Crop-raiding	Zhao (2005)
Stump-tailed macaque *M. arctoides*	Cambodia, China, India, Lao PDR, Malaysia, Myanmar, Thailand, Vietnam	V	V	Crop-raiding	Aggimarangsee (1992) and Hamada et al. (2007)

Long-tailed macaque *M. fascicularis*	Mainland SE Asia, Philippine and Indonesian archipelagos (introduced – various locales)	NT	LC	Tourism/provisioning	Fuentes and Gamerl (2005), Loudon et al. (2006), and Wheatley (1999)
				Urban nuisance	Malaivijitnond et al. (2011), Md-Zain et al. (2011), and Sha et al. (2009)
				Crop-raiding	Aggimarangsee (1992), Loudon et al. (2006), Marchal and Hill (2009), and Wheatley (2011)
Rhesus macaque *M. mulatta*	Afghanistan; Bangladesh, Bhutan, China, India, Lao PDR, Myanmar, Nepal, Pakistan, Thailand, Vietnam (introduced – various locales)	NT	LC	Tourism/provisioning	Jones-Engel et al. (2006) and Shek (2011)
				Urban nuisance	Chauhan and Pirta (2010a), Imam et al. (2002), Shek (2011), Southwick et al. (2005), Southwick and Siddiqi (2011), and Srivastava and Begum (2005)
				Crop-raiding	Chalise and Johnson (2005), Engeman et al. (2010), Pirta et al. (1997), Southwick et al. (2005), and Wang et al. (2006)
Japanese macaque *M. fuscata*	Japan	DD	LC	Tourism/provisioning	Knight (2005) and Watanabe and Muroyama (2005)
				Crop-raiding	Enari and Suzuki (2010), Knight (1999), Sprague (2002), Suzuki and Muroyama (2010), Watanabe and Muroyama (2005), and Yamada and Muroyama (2010)
Formosan macaque *M. cyclopis*	Taiwan	V	LC	Tourism/provisioning	Hsu et al. (2009)
				Crop-raiding	Poirier (1986)

[1] Listed IUCN categories in increasing order of concern: DD = Data Deficient; LC = Least Concern, NT = Near Threatened, V = Vulnerable, E = Endangered, CE = Critically Endangered (IUCN 2011). Species designated as V, E or CE are considered to be threatened with extinction.

Considering their wide geographical distribution and taxonomic diversity, the macaques are perhaps the most notorious and successful of 'pest primates'. All species raid crops (Table 14.1). Indeed, certain macaque species – the so-called weeds (Richard et al. 1989) – show a preference for foraging in the mosaic of habitats created by human settlement, cultivation and pastoralism and derive a substantial portion of their diet directly or indirectly from people (Richard et al. 1989). Unlike their 'pest' counterparts in Africa – the baboons and vervets – macaques have formed a commensal relationship with people in many Asian nations (Lane et al. 2010; Sha et al. 2009; Singh and Rao 2004; Southwick et al. 2005). Across Asia, macaques are found in proximity to villages and towns (Aggimarangsee 1992; Southwick et al. 1961; Watanabe and Muroyama 2005); some even make a living in densely populated urban areas (e.g. *M. mulatta* in Indian cities: Mathur and Manohar 1990; Srivastava and Begum 2005; *M. fascicularis* in residential Singapore: Lee and Chan 2011; Sha et al. 2009). This close association with people is facilitated by human cultural attitudes that imbue monkeys with religious and/or symbolic significance (Burton 2002; Knight 1999; Wheatley 1999; Wolfe 2002). For example, in Hindu mythology, monkeys are revered as representatives of Hanuman, the monkey god, following his key role in the Ramayana, a Hindu Sanskrit epic. Although Hanuman is usually depicted as a langur (*Semnopithecus entellus*), in many Hindu cultures, he has come to represent all monkeys, including macaques. Consequently, orthodox Hindus consider it their sacred duty to feed macaques (Pragatheesh 2011). More generally, macaques are commonly found in association with Hindu and Buddhist temples throughout south and southeast Asia and southern China, where they are provisioned by devotees and, at some sites, tourists (Aggimarangsee 1992; Jones-Engel et al. 2006; Loudon et al. 2006; Medhi et al. 2007; Southwick et al. 1961; Wheatley 1999; Zhao 2005). Whether the monkeys themselves are objects of worship or rather the sacred temples and shrines they often inhabit (Fuentes et al. 2005), cultural beliefs held in many parts of Asia have traditionally provided a context for tolerance and a measure of protection for macaque populations. Nevertheless, this close coexistence between humans and macaques inevitably leads to conflicts. Moreover, conflicts are increasingly challenging traditional relationships between people and macaques (Knight 1999; Southwick and Siddiqi 2011).

Conflicts between people and macaques occur in three broad contexts, all stemming from the macaques' dependence on humans for food, whether directly (i.e. provisioning) or indirectly (crop-raiding, food-stealing). First, macaques damage subsistence and/or cash crops in rural locales (Chakravarthy and Thyagaraj 2005; Chalise and Johnson 2005; Hashim et al. 2009; Priston 2005; Riley 2007; Supriatna et al. 1992; Suzuki and Muroyama 2010). Consequently, in agricultural areas, macaques may be viewed as serious vertebrate pests (Engeman et al. 2010; Knight 1999; Marchal and Hill 2009; Wang et al. 2006; Wheatley 2011). In rural Morocco, macaques damage commercially valuable timber by stripping the bark (Camperio Ciani et al. 2001). Second, macaques habituated to close interaction with people at temples and tourist attractions frequently show undesirable behaviours associated with provisioning, including human-directed aggression and food-snatching (Fa 1992; Fuentes and Gamerl 2005; Zhao 2005). Third, in urban towns and cities, macaques are sometimes regarded as a worrisome, potentially dangerous nuisance. Typical problem behaviours include physical aggression towards people, snatching bags, entering and damaging property, stealing food and other items, fouling and raiding garbage (Chauhan and Pirta 2010a; Cortes and Shaw 2006; Imam et al. 2002; Md-Zain et al. 2011; Sha et al. 2009; Shek 2011; Southwick et al. 2005). The 'monkey problem' may reach such proportions that urban macaques are regarded as a serious menace (Southwick and Siddiqi 2011; Southwick et al. 2005; Srivastava and Begum 2005). A further area of 'conflict' arising from close interaction between people and macaques, not covered in this review, concerns the potential for zoonotic disease transmission (see, e.g. Fuentes 2006a; Jones-Engel et al. 2006; Lane et al. 2010, and Chap. 10 by Peterson and Riley, this volume).

Below, we use specific studies to explore these aspects of human–macaque conflict in more detail and the influence of human cultural beliefs on attitudes towards macaques. The distinction between macaque behaviour in agricultural and urban settings, and at temples or other tourist attractions, is to some extent artificial – temple-residing monkeys raid farmers' subsistence crops (Aggimarangsee 1992; Loudon et al. 2006; Medhi et al. 2007), macaques provisioned at recreation sites can cause a nuisance in nearby residential areas (Cortes and Shaw 2006; Sha et al. 2009; Shek 2011), while macaques in rural locales may threaten people (Hamada et al. 2007; Knight 1999) and damage property (Enari and Suzuki 2010). Nevertheless, it is useful to consider human–macaque interactions in these environments individually to explore the variety of contexts in which conflict occurs.

14.2 Conflict Case Studies

14.2.1 Human–Macaque Conflict in Agricultural Areas

14.2.1.1 Crop-Raiding in Sulawesi, Indonesia

The island of Sulawesi in the Indonesian archipelago covers a relatively small area of 179,426 km^2 but supports a diversity of endemic species including the six to seven recognised species of Sulawesi macaque, representing a unique radiation within *Macaca* (Fooden 1980; see Riley 2010a for a recent review). All species are declining and are considered threatened by the International Union for Conservation of Nature with a noticeable recent decline indicated by a change in status from 2000 to 2008 (IUCN 2011) (Table 14.1). The chief threats are land conversion for agriculture, habitat fragmentation and hunting (Bynum et al. 1999; Lee 1995; IUCN 2011; Supriatna et al. 1992). Across Sulawesi, agriculture is one of the main economic activities, and much of the human population are engaged in subsistence agriculture (notably, sweet potato and maize) (Priston 2005; Whitten et al. 2002). However, recent years have seen an increase in cash-cropping, and many farmers now engage in wet-rice agriculture, and most also practise some form of plantation agriculture including coffee, cacao, cashew nut, palm oil and cloves (Riley and Priston 2010).

Fig. 14.1 Crop-raiding Buton macaque, Sulawesi (Courtesy: N. Priston)

Islam is the dominant religion in Sulawesi, but Christians form a substantial minority; Hindus and Buddhists are represented by small communities only. Earlier work identified crop-raiding as a critical obstacle to conservation of Sulawesi macaques (Bynum et al. 1999; Supriatna et al. 1992).

Current research on human–macaque interactions has focussed on two taxa, the Tonkean macaque *M. tonkeana* in central Sulawesi and the Buton macaque *M. brunnescens* on Buton Island in the southeast (synopsis in Riley and Priston 2010). These studies have considered crop-raiding as part of the macaque's ecological strategy in forest–farm ecotones and examined its impact on local livelihoods and the role of cultural attitudes in shaping local attitudes towards crop losses to monkeys (Priston 2005, 2009; Riley 2007, 2010b). Here, we focus principally on the Buton macaque. On Buton, local people engage predominantly in subsistence agriculture, with little cash-cropping; consequently, raiding of staple foods such as sweet potato, maize and banana is the primary concern in this region. These crops form an important component of the diet of some Buton macaques, with studied groups spending more than one-third of feeding time crop-raiding (Priston 2005) (Fig. 14.1). Nevertheless, sympatric wild pigs (*Sus* spp.) cause substantially more damage to sweet potato than monkeys (Priston 2009). Farmers' perceptions of loss were generally accurate, with farmers estimating an average loss of 9% of the farms' crops to monkeys at any one time versus actual measured losses of 10% (Priston 2005). This contrasts with the situation at Lore Lindu National Park where

Tonkean macaques were considered the most salient raiders of cacao, the region's major cash crop, despite causing considerably less damage than forest rats *Taeromys* sp. (Riley 2007)

Frequent damage to both subsistence and cash crops by macaques on Sulawesi, and the absence of a predominant Hindu or Buddhist culture that might be expected to foster tolerance of monkeys, predicts high levels of conflict between macaques and local people. In fact, currently, that appears not to be the case. For example, on Lore Lindu, traditional folklore envisions macaques as biologically and culturally related to humans, and harming crop-raiding macaques is traditionally taboo, lest the monkeys seek retribution (Riley 2010b). On Buton, there is little traditional folklore surrounding macaques. Nevertheless, the majority Muslim population is surprisingly tolerant of crop damage by monkeys, claiming that, though they dislike them, they do not wish to harm them (Priston 2005; Riley and Priston 2010). They describe macaques in human terms and express pity towards them as fellow creatures that also need food to survive. A commonly expressed sentiment is, 'if you don't want monkeys to come to your farm, don't open a farm', and losses to macaques are generally considered acceptable except in extreme cases (Priston 2005). Further, cashew farmers actually express positive attitudes towards crop-raiding macaques. The monkeys eat only the fruit, leaving the nuts scattered on the ground and easier for farmers to harvest (Priston 2005). In contrast, Hindus on the island (transmigrants from Bali) predominantly farm rice crops which are not raided by macaques. Despite this, and in spite of the traditional connection of monkeys to Hanuman in Hindu mythology, Buton's Hindus exhibit less tolerance of macaques compared to Muslims, and report hunting and eating them (Priston 2005; Riley and Priston 2010). These recent studies point to overall low current levels of human–macaque conflict on Sulawesi, despite frequent crop-raiding (Supriatna et al. 1992). Nevertheless, cultural attitudes towards wildlife and tolerance of crop losses are liable to change in accordance with shifting socio-economic and sociocultural conditions (see below). With continuing forest clearance and increased cash-cropping on Sulawesi, conflict between people and macaques is expected to increase.

14.2.1.2 Bark-Stripping in Morocco, North Africa

Barbary macaques (*M. sylvanus*) are the only members of the *Macaca* genus occurring naturally in the wild outside Asia. The species is popularly known from the introduced population on Gibraltar, which are visited by thousands of tourists annually (Fa and Lind 1996). However, the remaining wild *M. sylvanus* populations are limited to fragmented relict forests in mountainous or rocky areas in Morocco and Algeria. Surveys point to a sharp decline in Moroccan populations in recent decades associated with human degradation of the cedar and oak forests – the macaque's preferred habitat – as well as poaching for the illegal pet trade, predation by domestic dogs and drought (Camperio Ciani et al. 2005; Camperio Ciani and Mouna 2006; Mouna and Camperio Ciani 2006; van Lavieren 2008; van Lavieren and Wich 2010, also Chap. 11 by Majolo et al. this volume).

Unlike in Asia, local human cultures in North Africa do not attach religious sentiment to macaques.

The cedar oak forests of Morocco's Middle Atlas mountains represent the global stronghold for *M. sylvanus*. However, a conflict scenario has emerged between macaques and the human users of the forest. Macaques strip bark from commercially valuable cedar trees to satisfy water requirements and obtain certain nutrients (Camperio Ciani et al. 2001). Forest officials claim the density of monkeys in the Middle Atlas is increasing – contrary to research showing the population has declined (Camperio Ciani et al. 2005; van Lavieren and Wich 2010) – with resultant damage to the forest and, in particular, an increase in bark-stripping, which decreases the commercial value of the trees, occasionally causing their death. The monkeys are therefore considered economic pests, leading the Moroccan authorities to plan the relocation of macaques as a population management strategy (Mouna and Camperio Ciani 2006). Throughout the preceding decade, the Middle Atlas region was beset by drought, prompting shepherds to settle permanently near water sources. This had the effect of excluding monkeys from their natural water supply. Research indicates that the principal cause of the increase in bark-stripping by macaques stemmed from water shortage in response to drought and exacerbated by human occupation of water sources (Camperio Ciani et al. 2001). Despite clear evidence of the macaques' low density, and the detrimental impact of goats and sheep on the underbrush (e.g. reduced regeneration), many forest officials continue to espouse the viewpoint that macaque numbers are out of control (Mouna and Camperio Ciani 2006). Thus, endangered macaques have been made scapegoats for human-induced habitat degradation regionally.

14.2.2 Human–Macaque Conflict at Tourism Sites

Across Asia, at temples, nature parks, recreation sites and other tourist attractions, humans interact closely with macaques (*M. thibetana* in China: Matheson et al. 2006; Zhao 2005; *M. mulatta* in India and Nepal: Jones-Engel et al. 2006; Wolfe 2002; *M. fascicularis* in Indonesia and Thailand: Aggimarangsee 1992; Fuentes et al. 2005; *M. fuscata* in Japan: Knight 2005; *M. cyclopis* in Taiwan: Hsu et al. 2009) (Fig. 14.2). Elsewhere, free-ranging *M. sylvanus* are a long-established tourist attraction at Gibraltar in Europe (Fa and Lind 1996), and macaque viewing is currently being developed as a tourist attraction within the natural range of *M. sylvanus* in North Africa (e.g. Ifrane National Park, Morocco: Maréchal et al. 2011). While macaques lack religious connotations outside of Asia, a unifying characteristic of macaque-viewing sites is that the monkeys receive food from people. For religious devotees in Asia, feeding monkeys at sacred Hindu or Buddhist sites provides a means to obtain spiritual merit (Aggimarangsee 1992; Zhao 2005). Thus, macaques at tourist and/or religious sites are often conditioned to expect food from human visitors and therefore seek interactions with them. A small but growing literature concerns problems associated with provisioning free-ranging macaques at tourist

Fig. 14.2 Buddhist temple-dwelling rhesus macaque, Nepal (Courtesy: M. R. McLennan)

locations, particularly Gibraltar (e.g. Fa 1992; Fuentes 2006a,b) and the 'monkey temples' of Bali (Fuentes and Gamerl 2005; Fuentes 2006a; Lane et al. 2010; Loudon et al. 2006; Wheatley 1999; also Chap. 6 by Mallapur, this volume). Here, we use the example of Tibetan macaques in China to illustrate aspects of conflict that can arise from 'macaque tourism'.

14.2.2.1 Tourists, Temples and Tibetan Macaques: *M. thibetana* in China

Tibetan macaques and tourists interact at two sites in east-central China: Mt. Emei, a sacred Buddhist centre (Zhao 2005), and Mt. Huangshan, an ecotourism destination (Matheson et al. 2006). Both sites are popular attractions for domestic and international tourists. At Mt. Emei, pilgrims and tourists follow trails leading from the base of the mountain to the summit, along which they are intercepted by groups of macaques expecting food (Zhao 1999). Historically, limited feeding of macaques was performed by Buddhists monks and pilgrims, but following China's economic reform in 1980, growing numbers of visitors to Mt. Emei began feeding the monkeys for enjoyment. The macaques were thus conditioned to treat visitors as potential feeders (Zhao 2005). Arising from this scenario is one of the most extreme human–macaque conflicts thus far documented. *M. thibetana* is the largest and heaviest member of its genus, with adult male weighing as much as 22 kg (Zhao 1996). Their large body size makes these primates an intimidating and potentially dangerous adversary for humans. Macaques at Mt. Emei are fearless around visitors, routinely harassing them, for example, by manually inspecting clothes and baggage and 'robbing' them of food and possessions (Zhao and Deng 1992). This harassment can escalate into violent conflict, as when macaques chase, grab and bite humans, occasionally inflicting serious injury. A vicious circle has emerged: the monkeys are conditioned to use aggression to obtain food, while visitors use food as a 'passport' to aid safe passage (Zhao 2005). Extraordinarily, the deaths of ten visitors were indirectly attributed to macaques: six fell to their deaths retreating from threatening monkeys on narrow cliff paths, two others fell while trying to recover stolen possessions and a further two were killed by falling stones dislodged by macaques moving on cliffs above (Zhao and Deng 1992).

Macaques at Mt. Emei plainly distinguish visiting humans from locals, tending to target visitors who – unprepared for the monkeys' aggressive begging and robbing – typically adopt submissive behaviour (e.g. throwing food to placate the monkeys). Conversely, macaques show conditioned avoidance of local people who exhibit dominance behaviour towards them, for example, by using stones or other objects to drive them away. This defensive behaviour has had a deterrent effect, enabling locals to sell food without harassment from macaques, recover stolen property for tourists and escort frightened visitors past the monkeys (Zhao 2005). Despite these considerable problems, the management authority and travel agencies have been reluctant to highlight risks associated with feeding the macaques and educate visitors about appropriate behaviour, perhaps anticipating a reduction in entry fees (Zhao 2005).

At Mt. Huangshan, visitors are restricted to observation platforms and are officially prohibited from feeding macaques (Matheson et al. 2006). As a result, levels of aggression witnessed at Mt. Emei have largely been avoided at this site. Nevertheless, tourists' motivation to interact with monkeys frequently provokes aggressive threats from the macaques (McCarthy et al. 2009; Ruesto et al. 2010). For example, the most frequent macaque behaviours following pointing and rail-slapping – both common tourist behaviours – are facial threats and lunges/ground slaps, respectively (McCarthy et al. 2009). Tourist decibel levels on the viewing platform and the overall frequency of tourist behaviours – often of an attention-seeking nature – are associated with increased occurrence of macaque threats (Ruesto et al. 2010).

14.2.3 Human–Macaque Conflict in Urban Environments

14.2.3.1 Rhesus in Indian Towns and Cities

The rhesus macaque (*M. mulatta*) has the widest geographical distribution of the macaques and arguably the strongest commensal tendency. In many respects, it is the archetypal 'weed' macaque (Richard et al. 1989). In India, rhesus are found at temples, along roadsides and canals, in parks, rail stations, university campuses, in villages, towns and cities (Devi and Saikia 2008; Imam and Yahya 2002; Mathur and Manohar 1990; Medhi et al. 2007; Southwick et al. 1961, 2005). Surveys made by Southwick and colleagues since 1959 have revealed marked human-induced changes in the size of India's rhesus population and a shift in their habitat distribution. Between the late 1950s and 1970s, rhesus numbers declined by about 90% (Southwick and Siddiqi 1988). Contributing to this decline was a combination of anthropogenic factors: excessive trapping for biomedical research abroad, large-scale agricultural development, and weakening of cultural attitudes that traditionally afforded macaques protection from persecution (see below). Compared to rural areas, however, rhesus in towns fared better (Southwick and Siddiqi 1968). Rhesus numbers began to stabilise in the 1970s following reduced trapping and an eventual export ban in 1978. The 1970s also witnessed an increase in agricultural production and improved general economic conditions, which may have lessened animosity to rhesus compared with the 1960s when crop production was poor (Southwick et al. 2005). At any rate, local populations began to increase in some areas (Southwick et al. 1983; Southwick and Siddiqi 1988), and by the 1980s and 1990s, India's rhesus population showed signs of recovery (Southwick and Siddiqi 1988; Southwick et al. 2005).

Excluded from large parts of their former range by agricultural expansion and deforestation, rhesus monkeys have entered urban areas in increasing numbers (Southwick and Siddiqi 2011). Their successful colonisation of India's towns and cities was aided by human tolerance: unlike in rural areas, most urban residents are not farmers, and macaques do not compete directly with people for subsistence. Instead, the niche assumed by urban rhesus has been likened to that of rodents, pigeons or stray dogs; the monkeys often range around markets, bazaars and commercial areas living off food scraps, wastage and spillage (Southwick and Siddiqi 1968; Southwick et al. 1983). Additionally, some rhesus groups occupy temple grounds and leisure areas such as parks and receive food from visitors (Mathur and Manohar 1990). As discussed above, throughout the natural range of *Macaca*, humans feed macaques for pleasure. With competition over food relaxed, citizens of India's towns and cities are more inclined to share food with monkeys out of religious sentiment.

Nevertheless, recent studies indicate increased public concern over growing numbers of rhesus in India's urban centres (Chauhan and Pirta 2010a; Imam et al. 2002; Pirta et al. 1997; Southwick et al. 2005; Srivastava and Begum 2005). Escalating problems associated with macaques have been presented in the media or by management authorities as a 'monkey menace' (Chauhan and Pirta 2010b; Southwick and Siddiqi 2011). Surveys of public attitudes recount residents' feelings of harassment by macaques (Devi and Saikia 2008; Imam et al. 2002; Srivastava and Begum 2005). Common complaints are that rhesus invade homes and offices and steal food, clothes and other loose items; cause damage to roofs, television antennas and other electric wires; and 'vandalise' gardens. In Vrindaban town, near Agra City, residents complained that food could not be left unguarded or clothes dried in open areas. Money had to be spent fortifying homes and buildings with screens, iron grills and barbed wire to stop monkeys from entering (Imam et al. 2002; Southwick et al. 2005).

Moreover, rhesus macaques can be belligerent primates, and human-directed aggression is a potentially serious threat to human health and safety in urban centres. Aside from grabbing food, bags and other items – for example, snatching spectacles to obtain food in exchange (Chauhan and Pirta 2010a; Imam et al. 2002) – attacks on humans, including bites, are recorded regularly in some populations (Devi and Saikia 2008; Southwick et al. 2005). Most seriously, human-directed aggression by rhesus has been implicated in the deaths of several people in Indian cities. In circumstances paralleling the fatal interactions between tourists and Tibetan macaques at Mt. Emei (Zhao and Deng 1992), several people have fallen to their deaths from rooftops during aggressive altercations with rhesus (Southwick et al. 2005; Southwick and Siddiqi 2011). In an especially newsworthy incident, the deputy major of Delhi died after falling from his roof fending off monkeys (BBC News 2007). As in other contexts where people and macaques interface (e.g. at tourist attractions), an additional concern arising from close human–macaque contact in urban environments – particularly monkey bites – is the potential for disease transmission (e.g. herpes B virus; see Jones-Engel et al. 2006; Southwick et al. 2005).

The ecological and spiritual relationship that people traditionally had with macaques in India has undergone change. Whereas religious taboos previously fostered tolerance, in today's rapidly modernising India, people are more pragmatic. Managers of large-scale agricultural developments have no sentimental attachments to rhesus (Southwick et al. 1983). Damage to personal property and other nuisance behaviour by urban monkeys – in particular, occasional incidents of severe aggression – fosters unsympathetic attitudes among citizens of India's towns and cities. Revealingly, many Hindus in Guwahati City, Assam, expressed no religious sentiment towards rhesus (Srivastava and Begum 2005). Southwick and Siddiqi (2011:288) point out that the rhesus macaque's transition from revered representatives of Hanuman to economic and public health pests has produced 'cultural and philosophical conflicts for many people of India'. While some Hindus continue to feel uncomfortable about translocation or culling as management options for troublesome macaques (Chauhan and Pirta 2010b), others are calling for solutions to their 'monkey problem' (see below). The worsening relationship between people and India's most commensal of primates has wider implications. More than half of India's primates are rare and endangered (Southwick and Siddiqi 2001). There are fears that the escalating conflict between rhesus and people will erode public support for primate conservation in general (Imam et al. 2002; Southwick and Siddiqi 2001).

It is not only in India where paradoxical attitudes towards macaques prevail or where 'cultural and psychological conflicts' have emerged out of a shift in human–macaque ecological interactions (see, e.g. Knight's (1999) cultural analysis of the growing conflict between people and macaques *M. fuscata* in rural Japan). People's perceptions of, and attitudes towards, wildlife are not fixed in time or space. Across Asia, cultural beliefs that traditionally fostered tolerance of macaques are weakening in rapidly modernising societies. As farmers become absorbed into a market economy, losses to macaques, which previously may have been accepted as part of general crop returns, assume a far greater perceptual importance (Lee and Priston 2005; Southwick and Siddiqi 2011). Increasingly, tolerance of macaques may have more to do with their economic value as tourist attractions and less to do with traditional religious sentiments (Schilaci et al. 2010). Changing sociocultural circumstances can result in relaxation of taboos. For example, while traditional indigenous folklore fosters human–macaque coexistence in central Sulawesi (Riley 2010b), migrants have no such taboos. This implies that increased transmigration will have negative consequences for Sulawesi macaques (Riley 2007). Further, the lower tolerance shown by migrant Balinese Hindus towards macaques compared to Muslims on Buton Island (Priston 2005; Riley and Priston 2010) illustrates that Hindus are not necessarily more accommodating of macaques than followers of religions that do not mythologise monkeys.

14.3 Conflict Management

The rise in human–macaque conflicts throughout the natural range of *Macaca*, and in regions where macaques have been introduced (Engeman et al. 2010; Wheatley 2011), present substantial challenges to the sustainability of human–macaque relationships and call for effective management strategies to facilitate coexistence. Since conflicts occur in a variety of settings (including farms, villages, temples, recreational areas, town and cities) and contexts (resource overlap, crop-raiding, food provisioning, tourism, commensalism), multiple management strategies are needed. Specific conflicts may require an integrated approach with interventions tailored to local situations (e.g. Shek 2011). A thorough discussion of the relative advantages and disadvantages of various conflict management approaches is beyond the scope of this chapter (for a comprehensive overview of human–macaque conflict mitigation strategies, see Jones-Engel et al. 2011). However, it is important to emphasise that conflict mitigation should not be considered in isolation from other factors affecting the sustainability of macaque populations (e.g. habitat destruction, loss of natural food source, hunting) but as part of an integrated conservation strategy. Here, we discuss several conflict management approaches which fall into two broad categories: (1) interventions that aim to alter the behaviour of macaques and/or people and (2) interventions that seek to control the size, demography or distribution of macaque populations.

14.3.1 Behavioural Management

14.3.1.1 Crop-Raiding

Macaques that are considered a significant threat to local livelihoods are unlikely to be viewed as a resource to conserve. Thus, development of non-lethal strategies to alleviate crop damage by macaques is imperative. Once established, however, crop-raiding behaviour in cognitively complex animals such as primates can be extremely difficult to change (Chakravarthy and Thyagaraj 2005; Hill 2005; Strum 1994), and no single method is wholly effective at preventing monkey raids. Among commonly used techniques, vigilant guarding and chasing off intruding monkeys (using dogs, slingshots, firecrackers and so on) can be a successful deterrent but is costly in time and labour (Hill 2005). On the other hand, crop damage to cacao by Tonkean macaques in Sulawesi was independent of guarding frequency (Riley 2007). Traditional fences (e.g. made of thorns or branches) are largely ineffective against raiding primates, which can usually navigate around them owing to their extreme agility (Wang et al. 2006). In Japan, specially designed electric fences effectively exclude large to medium crop-raiding wildlife, including macaques *M. fuscata* (Honda et al. 2009).[1] However, their high cost makes them impractical in most rural localities in Asia and North Africa where macaque crop-raiding is a problem. Experimental studies of taste aversion in baboons have produced promising results (Forthman et al. 2005) but appear not to have been replicated with wild crop-raiding macaques. Chakravarthy and Thyagaraj (2005) reported that trimming and debranching of shade trees significantly reduced damage to cardamom plantations by bonnet macaques, *M. radiata*.

An alternative or complementary strategy for farmers is to plant alternative, buffer crops which are unattractive to monkeys in high-conflict zones such as along the edges of macaque habitat. For example, coffee is the second most important cash crop for farmers in central Sulawesi and is not raided by macaques (Riley 2007). However, promotion of buffer crops requires consideration of relative market prices and harvesting costs between crops. Farmers are unlikely to switch to alternative crops if it involves economic or labour costs (Riley and Priston 2010). A fuller understanding of conflicts associated with crop damage in rural landscapes requires studies of the ecology of macaque crop-raiding (Yamada and Muroyama 2010). When human activities reduce natural foods to the extent that they are insufficient to support resident macaques, population management strategies are needed to resolve conflicts over resources (see below).

14.3.1.2 Macaque Tourism

Studies of tourist–macaque interactions in China and elsewhere where humans interact with free-ranging macaques (e.g. Gibraltar and Bali) clearly indicate that reduced opportunities for physical contact between visitors and macaques and greater regulation of visitor behaviour are necessary to promote a more positive tourist experience at macaque-viewing sites. At Mt. Huangshan, the second of the

[1]The Japan situation is notable because, unlike many other parts of Asia, increased acculturation of macaques to anthropogenic environments and an associated rise in crop-raiding is not linked to human population growth in the macaques' range. Rather, a significant human depopulation of rural areas has occurred since the 1950s (Watanabe and Muroyama 2005). Reduced human presence on the land has compromised farmers' ability to protect their farmland. However, changes in land use, particularly the replacement of natural vegetation with conifer plantations, are also likely contributory factors in the escalation of conflict (Agetsuma 2007; Watanabe and Muroyama 2005).

sites where Tibetan macaques and tourists interact, ecotourism was developed intentionally to avoid problems experienced at Mt. Emei. Visitors are restricted to viewing platforms, thereby minimising physical proximity, and provisioning is performed by trained staff at scheduled times (Matheson et al. 2006). These measures have substantially reduced aggressive interactions between people and macaques relative to Mt. Emei (Zhao 2005). Even so, there and at other macaque-viewing sites, stronger regulations need to be put in place to discourage visitor behaviours that provoke aggressive responses from macaques (Fuentes and Gamerl 2005; McCarthy et al. 2009; Zhao 2005), including noise regulation (Ruesto et al. 2010). Thus, there is a parallel need for improved public education about macaque behaviour and appropriate human behaviour at interaction sites, for example, through prominent signage, staff talks and visitor centres (Lee and Chan 2011; Shek 2011). Increasingly, such information is available to visitors at macaque-viewing sites to varying extents, but a major problem is that tourists may disregard it. For example, while public feeding of macaques is officially prohibited at Mt. Huangshan, surreptitious provisioning occurs nonetheless (Ruesto et al. 2010). In Gibraltar, frequent unofficial feeding by tourists, tour guides and taxi drivers is the main reason for physical interaction between people and Barbary macaques (Fuentes 2006b), despite laws and public campaigns to address the problem (Cortes and Shaw 2006). Even so, public education programmes combined with proper law enforcement can be effective in limiting undesirable human behaviours including unofficial feeding, thereby substantially reducing negative human–macaque interactions as well as the potential for pathogen transmission. In Singapore, auxiliary police and surveillance cameras are used to enforce a no-feeding law. The fine for feeding long-tailed macaques was recently increased to SGD$500 (Lee and Chan 2011; Sha et al. 2009). However, such stringent measures are unlikely to be feasible in all macaque-viewing contexts (e.g. rural temples, roadsides). An additional problem is that stricter regulations are likely to be unpopular with some local people such as vendors who sell monkey feed to visitors (Zhao 2005) and tour guides who use food to lure monkeys to interact with tourists (Fuentes 2006b). Given the important role of macaque viewing in providing financial incentives for local people to support conservation efforts, care must be taken to avoid alienating local communities.

Education efforts can also help ameliorate conflicts in urban and residential areas. Since most human–macaque conflicts are food-related, practical interventions to reduce nuisance behaviours include limiting macaques' access to refuge, for example, through replacing rubbish bins with animal-proof bins and managing refuge collection so as to remove food sources (Jones-Engel et al. 2011; Md-Zain et al. 2011; Shek 2011). Fortification of buildings with protective barriers such as screens, bars and wire or electric fences will help prevent monkey incursions into homes and offices but may not be fully effective; long-tailed macaques in urban Thailand have been observed climbing over such barriers (Malaivijitnond et al. 2011). In Hong Kong, food trees were planted within country parks to discourage macaques from straying into nearby residential areas in search of food (Shek 2011). Similarly, fruit trees along residential streets in Singapore were replaced by trees that are less attractive as food for macaques (Lee and Chan 2011).

14.3.2 Population Management

Solutions to human–macaque conflicts sometimes necessitate population management through control of the size, demography or distribution of macaque populations. Official culling of macaques as a method to address problem behaviour has been carried out in several Asian range countries (e.g. *M. fascicularis* in Singapore: Sha et al. 2009; *M. fuscata* in Japan: Knight 1999), in Gibraltar (Fa and Lind 1996; Cortes and Shaw 2006) and in an attempt to eradicate introduced *M. fascicularis* on Ngeaur Island, Republic of Palau (Wheatley 2011). Culling was recently proposed as a strategy to eliminate invasive *M. mulatta* in Puerto Rico (Engeman et al. 2010). However, culling may be an undesirable management strategy both for conservation and cultural reasons – for example, it is unacceptable to devout Hindus in parts of Asia (Chauhan and Pirta 2010b) – and other methods to manage populations should be considered in most situations.

14.3.2.1 Translocation

Translocation (or relocation) of problem macaques has been employed as a non-lethal solution to human–macaque conflicts, particularly concerning urbanised rhesus in India (Imam et al. 2001, 2002; Imam and Yahya 2002). In one intervention, 600 monkeys were successfully translocated from Vrindaban, where conflict had reached high levels, to eight semi-forested sites deemed to have adequate natural food, water and shelter for macaques. (In fact, release sites were located in areas where rhesus were formerly present but had been extirpated following intensive trapping in the 1950s and 1960s; Southwick et al. 2005). Four years following this intervention, all troops remained at their release sites, apparently accepted by local people (Imam et al. 2002). However, in addition to the potential high costs involved, among other concerns (see Massei et al. 2010), there are some situations where translocation is inappropriate – for example, when there is a lack of suitable habitat to move animals because of extensive habitat modification by people (Srivastava and Begum 2005). In forest–farm ecotones removal (or elimination) of macaques may not necessarily end a crop problem due to immigration of monkeys from adjacent areas (Chakravarthy and Thyagaraj 2005). Further, translocation may simply spread the 'monkey menace' from one place to another. One relocated rhesus troop increased from 20 to 258 individuals in 25 years, potentially stretching local people's tolerance to the limit (Southwick and Siddiqi 2011). Thus, in some circumstances, translocation may be ineffective at controlling macaque populations.

The decision to relocate wild macaques must be based on a sound appraisal of the relative costs and benefits of such a management approach and a good understanding of the causes of a conflict situation. In Morocco's Middle Atlas, translocation of endangered *M. sylvanus* was deemed necessary by forest authorities based on the erroneous assertion that macaque damage to commercially valuable timber stocks (bark-stripping) was due to overpopulation. In fact, macaques in the region are declining (Camperio Ciani and Mouna 2006). As discussed above, the main factor eliciting bark-stripping was lack of access to water due to settlement around water sources by shepherds. Consequently, the problem could potentially be alleviated by enabling the macaque's access to water sources – a likely more effective and economical strategy than large-scale population relocation (Camperio Ciani et al. 2001).

14.3.2.2 Fertility Control

Sterilisation and/or contraceptive programmes represent alternative management tools that are increasingly proposed as effective means of controlling macaque populations, particularly in urban settings (Malaivijitnond et al. 2011; Rattan 2011; Shek 2011). In Hong Kong, a large-scale contraceptive and neutering programme has been running since 2007 in an attempt to limit the expanding monkey population (Shek 2011). After trapping, female macaques are injected with an immunocontraceptive vaccine while males are vasectomised. As of 2010, 50–60% of Hong Kong's macaques have been treated, and initial results indicate a decline in the total population (Shek 2011). Fertility control is an attractive management option for reducing human–macaque conflict because it avoids directly killing animals as well as the various costs and problems associated with translocation. Moreover, where local people are uncomfortable with culling or removal of macaques out of religious sentiment, sterilisation programmes are likely to gain public support (Chauhan and Pirta 2010b). However, as Jones-Engel et al. (2011) point out, long-term studies of the effects of sterilisation/contraceptive programmes on populations and individuals (including behavioural consequences), and the efficiency of this approach to mitigate human–macaque conflict, are presently lacking. Clearly, systematic investigation in this regard is needed.

Finally, Southwick and Siddiqi (2011) suggest that moderate harvest for legitimate biomedical research represents an option for controlling India's growing rhesus population, potentially to the benefit of both people and macaques. They argue that the decision in 1978 to halt all trapping for export was short-sighted because it failed to consider the rhesus monkey's high reproductive rates and commensal tendency and the ongoing destruction of their natural habitats. They acknowledge, however, that cultural and religious factors may preclude the sustainable harvesting of rhesus as a population management and conflict mitigation strategy.

14.4 Conclusions

Macaque–human relationships are complex and culturally specific, ranging from relatively peaceful coexistence to extreme levels of conflict. Of all primates, macaques adjust particularly well to human-modified environments, both rural and urban, and in some contexts, develop commensal, mutually beneficial relationships with humans. In this chapter, we focussed on three main areas of human–macaque interaction: rural crop damage, tourism sites and urban contexts. Despite high levels of potential conflict, there are remarkable levels of human tolerance to macaques in these contexts, demonstrated either through deliberate provisioning, seeking out of interactions (in the tourist context) or relative indifference (e.g. Buton). This provides hope for the conservation of macaques. For every example of tolerance, however, there are numerous instances of conflict, and with ever-expanding human populations, increased movement of people within and between countries, and changing socio-economic conditions, the future of human–macaque relationships is likely to be an uneasy one. As illustrated in India, changing cultural attitudes can change traditional human–macaque relationships, usually detrimentally for macaques. Macaque conservation requires a multiple management strategy approach depending on the specific context, and no single management strategy will suit all sites of human–macaque interaction. Conservation strategies should focus on promotion of tolerant cultural attitudes in addition to reduction of negative interactions in order to ensure long-term survival of macaque populations.

References

Abegg C (2006) The role of contingency in the evolution of the Barbary macaque. In: Hodges JK, Cortes J (eds) The Barbary macaque: biology, management and conservation. Nottingham University Press, Nottingham, pp 17–27

Agetsuma N (2007) Ecological function losses caused by monotonous land use induce crop raiding by wildlife on the island of Yakushima, southern Japan. Ecol Res 22:390–402

Aggimarangsee N (1992) Survey for semi-tame colonies of macaques in Thailand. Nat Hist Bull Siam Soc 40:103–166

BBC News (2007) Monkeys attack Delhi politician. 21 Oct 2007 http://news.bbc.co.uk/1/hi/world/south_asia/7055625.stm. Accessed Aug 2011

Burton FD (2002) Monkey King in China: basis for a conservation policy? In: Fuentes A, Wolfe LD (eds) Primates face to face: the conservation implications of human–nonhuman primate interconnections. Cambridge University Press, Cambridge, pp 137–162

Bynum EL, Kohlhaas AK, Pramono AH (1999) Conservation status of Sulawesi macaques. Trop Biodivers 6:123–144

Camperio Ciani A, Mouna M (2006) Human and environmental causes of the rapid decline of *Macaca sylvanus* in the Middle Atlas of Morocco. In: Hodges JK, Cortes J (eds) The Barbary macaque: biology, management and conservation. Nottingham University Press, Nottingham, pp 257–275

Camperio Ciani A, Martinoli L, Capiluppi C, Arahou M, Mouna M (2001) Effects of water availability and habitat quality on bark-stripping behavior in Barbary macaques. Conserv Biol 15:259–265

Camperio Ciani A, Palentini L, Arahou M, Martinoli L, Capiluppi C, Mouna M (2005) Population decline of Macaca sylvanus in the Middle Atlas of Morocco. Biol Conserv 121:635–641

Chakravarthy AK, Thyagaraj NE (2005) Coexistence of bonnet macaques (*Macaca radiata radiata* Geoffroy) with planters in the cardamom (*Elettaria cardamomum* Maton) and coffee (*Coffea arabica* Linnaeus) plantations of Karnataka, South India: hospitable or hostile? In: Paterson JD, Wallis J (eds) Commensalism and conflict: the human–primate interface. American Society of Primatologists, Norman, pp 271–293

Chalise MK, Johnson RL (2005) Farmer attitudes toward the conservation of "pest" monkeys: the view from Nepal. In: Paterson JD, Wallis J (eds) Commensalism and conflict: the human–primate interface. American Society of Primatologists, Norman, pp 223–239

Chapman CA, Peres CA (2001) Primate conservation in the new millennium: the role of scientists. Evol Anthropol 10:16–33

Chauhan A, Pirta RS (2010a) Agonistic interactions between humans and two species of monkeys (rhesus monkey *Macaca mulatta* and hanuman langur *Semnopithecus entellus*) in Shimla, Himachal Pradesh. J Psychol 1:9–14

Chauhan A, Pirta RS (2010b) Public opinion regarding human-monkey conflict in Shimla, Himachal Pradesh. J Hum Ecol 30:105–109

Choudhury A (2003) The Pig-tailed Macaque *Macaca nemestrina* in India – status and conservation. Primate Conserv 19:91–98

Cortes J, Shaw E (2006) The Gibraltar macaques: management and future. In: Hodges JK, Cortes J (eds) The Barbary macaque: biology, management and conservation. Nottingham University Press, Nottingham, pp 199–210

Cowlishaw G, Dunbar R (2000) Primate conservation biology. University of Chicago Press, Chicago

de Freitas CH, Setz EZF, Araújo ARB, Gobbi N (2008) Agricultural crops in the diet of bearded capuchin monkeys, *Cebus libidinous Spix* (Primates: Cebidae), in forest fragments in southeast Brazil. Revista Brasileira de Zoologia 25:32–39

Deag JM (1977) The status of the Barbary macaque *Macaca sylvanus* in captivity and factors affecting its distribution in the wild. In: Rainier HSH, Bourne GH (eds) Studies in primate conservation. Academic, New York, pp 267–287

Devi OS, Saikia PK (2008) Human–monkey conflict: a case study at Gauhati University Campus, Jalukbari, Kamrup, Assam. Zoos' Print xxiii:15–18

Else JG (1991) Nonhuman primates as pests. In: Box HO (ed) Primate responses to environmental change. Chapman & Hall, London, pp 155–165

Enari H, Suzuki T (2010) Risk of agricultural and property damage associated with the recovery of Japanese monkey populations. Landsc Urban Plann 97:83–91

Engeman RM, Laborde JE, Constantin BU, Shwiff SA, Hall P, Duffiney A, Luciano F (2010) The economic impacts to commercial farms from invasive monkeys in Puerto Rico. Crop Prot 29:401–405

Fa JE (1992) Visitor-directed aggression among the Gibraltar macaques. Zoo Biol 11:43–52

Fa JE, Lind R (1996) Population management and viability of the Gibraltar Barbary macaques. In: Fa JE, Lindburg DG (eds) Evolution and ecology of macaque societies. Press Syndicate of the University of Cambridge, Cambridge, pp 235–262

Fooden J (1980) Classification and distribution of living macaques. In: Lindburg DG (ed) The macaques: studies in ecology behavior and evolution. Van Nostrand Reinhold Co., New York, pp 1–9

Forthman DL, Strum SC, Muchemi G (2005) Applied conditioned taste aversion and the management and conservation of crop-raiding primates. In: Paterson JD, Wallis J (eds) Commensalism and conflict: the human–primate interface. American Society of Primatologists, Norman, pp 420–443

Forthman-Quick DL, Demment M (1988) Dynamics of exploitation: differential energetic adaptations of two troops of baboons to recent human contact. In: Fa JE, Southwick C (eds) Ecology and behaviour of food enhanced primate groups. Liss, New York, pp 25–51

Fuentes A (2002) Monkeys, humans and politics in the Mentawai Islands: no simple solutions in a complex world. In: Fuentes A, Wolfe LD (eds) Primates face to face: the conservation implications of human–nonhuman primate interconnections. Cambridge University Press, Cambridge, pp 187–207

Fuentes A (2006a) Human culture and monkey behavior: assessing the contexts of potential pathogen transmission between macaques and humans. Am J Primatol 68:880–896

Fuentes A (2006b) Patterns and context of human–macaque interactions in Gibraltar. In: Hodges JK, Cortes J (eds) The Barbary macaque: biology, management and conservation. Nottingham University Press, Nottingham, pp 169–184

Fuentes A, Gamerl S (2005) Disproportionate participation by age/sex classes in aggressive interactions between long-tailed macaques (*Macaca fascicularis*) and human tourists at Padangtegal Monkey Forest, Bali, Indonesia. Am J Primatol 66:197–204

Fuentes A, Southern M, Suaryana KG (2005) Monkey forests and human landscapes: is extensive sympatry sustainable for *Homo sapiens* and *Macaca fascicularis* on Bali? In: Paterson JD, Wallis J (eds) Commensalism and conflict: the human–primate interface. American Society of Primatologists, Norman, pp 168–195

Gautier J-P, Biquand S (1994) Primate commensalism. Revue d'Ecologie (La Terre et la Vie) 49:210–211

Groves CP (2001) Primate taxonomy. Smithsonian Institution Press, Washington

Hamada Y, Malaivijitnond S, Kingsada P, Bounnam P (2007) The distribution and present status of primates in the northern region of Lao PDR. Nat Hist J Chulalongkorn Univ 7:161–191

Hashim NR, Abdul Manan MS, Nazli MF (2009) Using geographic information system to predict primate crop raiding in Peninsular Malaysia. IUP J Environ Sci 3(4):39–46

Hill CM (2005) People, crops, and primates: a conflict of interests. In: Paterson JD, Wallis J (eds) Commensalism and conflict: the human–primate interface. American Society of Primatologists, Norman, pp 41–59

Hill C, Osborn F, Plumptre AJ (2002) Human–wildlife conflict: identifying the problem and possible solutions, vol 1, Albertine Rift Technical Report Series. Wildlife Conservation Society, New York

Honda T, Miyagawa Y, Ueda H, Inoue M (2009) Effectiveness of newly-designed electric fences in reducing crop damage by medium and large mammals. Mamm Study 34:13–17

Hsu MJ, Kao C-C, Agoramoorthy A (2009) Interactions between visitors and Formosan macaques (Macaca cyclopis) at Shou-Shan Nature Park, Taiwan. Am J Primatol 71:214–222

Imam E, Yahya HSA (2002) Management of monkey problem in Aligarh Muslim University, Uttar Pradesh. Zoos' Print J 17:685–687

Imam E, Malik I, Yahya HSA (2001) Translocation of rhesus macaques from Airforce Station, Gurgaon (Haryana) to the natural forest of Ferozpur-Jhirka, Haryana, India. J Bombay Nat Hist Soc 98:355–359

Imam E, Yahya HSA, Malik I (2002) A successful mass translocation of commensal rhesus monkeys *Macaca mulatta* in Vrindaban, India. Oryx 36:87–93

IUCN (2011) IUCN red list of threatened species. Version 2011.1. http://www.iucnredlist.org. Accessed 7 Aug 2011

Jones-Engel L, Engel GA, Heidrich J, Chalise M, Poudel N, Viscidi R, Barry PA, Allan JS, Grant R, Kyes R (2006) Temple monkeys and health implications of commensalism, Kathmandu, Nepal. Emerg Infect Dis 12:900–906

Jones-Engel L, Engel G, Gumert MD, Fuentes A (2011) Developing sustainable human–macaque communities. In: Gumert MD, Fuentes A, Jones-Engel L (eds) Monkeys on the edge: ecology and management of long-tailed macaques and their interface with humans. Cambridge University Press, Cambridge, pp 295–327

Knight J (1999) Monkeys on the move: the natural symbolism of human–macaque conflict in Japan. J Asian Stud 58:622–647

Knight J (2005) Feeding Mr Monkey: cross-species food 'exchange' in Japanese monkey parks. In: Knight J (ed) Animals in person: cultural perspectives on human–animal intimacy. Berg, Oxford, pp 231–253

LaFleur M, Gould L (2009) Feeding outside the forest: the importance of crop raiding and an invasive weed in the diet of gallery forest ring-tailed lemurs (*Lemur catta*) following a cyclone at the Beza Mahafaly Special Reserve, Madagascar. Folia Primatol 80:233–246

Lane KE, Lute M, Rompis A, Wandia IN, Putra IGAA, Hollocher H, Fuentes A (2010) Pests, pestilence, and people: the long-tailed macaque and its role in the cultural complexities of Bali. In: Gursky-Doyen S, Supriatna J (eds) Indonesian primates. Springer, New York, pp 235–248

Lee RJ (1995) Population survey of the crested black macaque (*Macaca nigra*) at Manembonembo Nature Reserve in North Sulawesi, Indonesia. Primate Conserv 16:63–65

Lee BPY-H, Chan S (2011) Lessons and challenges in the management of long-tailed macaques in urban Singapore. In: Gumert MD, Fuentes A, Jones-Engel L (eds) Monkeys on the edge: ecology and management of long-tailed macaques and their interface with humans. Cambridge University Press, Cambridge, pp 307–313

Lee PC, Priston NEC (2005) Human attitudes to primates: perceptions of pests, conflict and consequences for primate conservation. In: Paterson JD, Wallis J (eds) Commensalism and conflict: the human–primate interface. American Society of Primatologists, Norman, pp 1–23

Linkie M, Dinata Y, Nofrianto A, Leader-Williams N (2007) Patterns and perceptions of wildlife crop raiding in and around Kerinci Seblat National Park, Sumatra. Anim Conserv 10: 127–135

Loudon JE, Howells ME, Fuentes A (2006) The importance of integrative anthropology: a preliminary investigation employing primatological and cultural anthropological data collection methods in assessing human-monkey co-existence in Bali, Indonesia. Ecol Environ Anthropol 2:2–13

Malaivijitnond S, Vazquez Y, Hamada Y (2011) Human impact on long-tailed macaques in Thailand. In: Gumert MD, Fuentes A, Jones-Engel L (eds) Monkeys on the edge: ecology and management of long-tailed macaques and their interface with humans. Cambridge University Press, Cambridge, pp 118–156

Marchal V, Hill C (2009) Primate crop-raiding: a study of local perceptions in four villages in North Sumatra, Indonesia. Primate Conserv 24:107–116

Maréchal L, Semple S, Majolo B, Qarro M, Heistermann M, MacLarnon A (2011) Impacts of tourism on anxiety and physiological stress levels in wild male Barbary macaques. Biol Conserv 144:2188–2193

Massei G, Quy RJ, Gurney J, Cowan DP (2010) Can translocations be used to mitigate human–wildlife conflicts? Wildl Res 37:428–439

Matheson MD, Sheeran LK, Li J-H, Wagner RS (2006) Tourist impact on Tibetan macaques. Anthrozoös 19:158–168

Mathur R, Manohar BR (1990) Density of *Macaca mulatta* and *Presbytis entellus* in the old city of Jaipur: a three year survey. Appl Anim Behav Sci 27:351–361

McCarthy MS, Matheson MD, Lester JD, Sheeran LK, Li J-H, Wagner RS (2009) Sequences of Tibetan macaque (*Macaca thibetana*) and tourist behaviors at Mt. Huangshan, China. Primate Conserv 24:145–151

McLennan MR (2008) Beleaguered chimpanzees in the agricultural district of Hoima, western Uganda. Primate Conserv 23:45–54

Md-Zain BM, Tarmizi MR, Mohd-Zaki M (2011) Campus monkeys of Universiti Kebangsaan Malaysia: nuisance problems and students' perceptions. In: Gumert MD, Fuentes A, Jones-Engel L (eds) Monkeys on the edge: ecology and management of long-tailed macaques and their interface with humans. Cambridge University Press, Cambridge, pp 101–117

Medhi R, Chetry D, Basavdatta C, Bhattacharjee PC (2007) Status and diversity of temple primates in northeast India. Primate Conserv 22:135–138

Menon S, Poirier FE (1996) Lion-tailed macaques (*Macaca silenus*) in a disturbed forest fragment: activity patterns and time budget. Int J Primatol 17:969–985

Mouna M, Camperio Ciani A (2006) Distribution and demography of the Barbary macaque (*Macaca sylvanus* L.) in the wild. In: Hodges JK, Cortes J (eds) The Barbary macaque: biology, management and conservation. Nottingham University Press, Nottingham, pp 239–255

Nahallage CAD, Huffman MA, Kuruppu N, Weerasingha T (2008) Diurnal primates in Sri Lanka and people's perception of them. Primate Conserv 23:81–87

Nijman V, Nekaris KAI (2010) Testing a model for predicting primate crop-raiding using crop- and farm-specific risk values. Appl Anim Behav Sci 127:125–129

Pirta RS, Gadgil M, Kharshikar AV (1997) Management of the rhesus monkey *Macaca mulatta* and Hanuman langur *Presbytis entellus* in Himachal Pradesh, India. Biol Conserv 79:97–106

Poirier FE (1986) A preliminary study of the Taiwan macaque (*Macaca cyclopis*). Zoolog Res 7:411–422

Pragatheesh A (2011) Effect of human feeding on the road mortality of rhesus macaques on National Highway–7 routed along Pench Tiger Reserve, Madhya Pradesh, India. J Threat Taxa 3:1656–1662

Priston NEC (2005) Crop-raiding by Macaca ochreata brunnescens in Sulawesi: reality, perceptions and outcomes for conservation. PhD thesis. Cambridge University Press, Cambridge
Priston NEC (2009) Exclosure plots as a mechanism for quantifying damage to crops by primates. Int J Pest Manag 55:243–249
Rattan SK (2011) Managing human–macaque conflict in Himachal, India. In: Gumert MD, Fuentes A, Jones-Engel L (eds) Monkeys on the edge: ecology and management of long-tailed macaques and their interface with humans. Cambridge University Press, Cambridge, pp 283–285
Richard AF, Goldstein SJ, Dewar RE (1989) Weed macaques: the evolutionary implications of macaque feeding ecology. Int J Primatol 10:569–594
Riley EP (2007) The human–macaque interface: conservation implications of current and future overlap and conflict in Lore Lindu National Park, Sulawesi, Indonesia. Am Anthropol 109:473–484
Riley EP (2010a) The endemic seven: four decades of research on the Sulawesi macaques. Evolut Anthropol 19:22–36
Riley EP (2010b) The importance of human–macaque folklore for conservation in Lore Lindu National Park, Sulawesi, Indonesia. Oryx 44:235–240
Riley EP, Priston NEC (2010) Macaques in farms and folklore: exploring the human–nonhuman primate interface in Sulawesi, Indonesia. Am J Primatol 72:848–854
Riley EP, Suryobroto B, Maestripieri D (2007) Distribution of *Macaca ochreata* and identification of mixed ochreata-tonkeana groups in south Sulawesi, Indonesia. Primate Conserv 22:129–133
Ruesto LA, Sheeran LK, Matheson MD, Li J-H, Wagner RS (2010) Tourist behavior and decibel levels correlate with threat frequency in Tibetan macaques (*Macaca thibetana*) at Mt. Huangshan, China. Primate Conserv 25:99–104
Schilaci MA, Engel GA, Fuentes A, Rompis A, Putra A, Wandia IN, Bailey JA, Brogdon BG, Jones-Engel L (2010) The not-so-sacred monkeys of Bali: a radiographic study of human–primate commensalism. In: Gursky-Doyen S, Supriatna J (eds) Indonesian primates. Springer, New York, pp 249–256
Sha JCM, Gumert MD, Lee BPY-H, Jones-Engel L, Chan S, Fuentes A (2009) Macaque–human interactions and the societal perceptions of macaques in Singapore. Am J Primatol 71:825–839
Shek CT (2011) Management of nuisance macaques in Hong Kong. In: Gumert MD, Fuentes A, Jones-Engel L (eds) Monkeys on the edge: ecology and management of long-tailed macaques and their interface with humans. Cambridge University Press, Cambridge, pp 297–301
Singh M, Rao NR (2004) Population dynamics and conservation of commensal bonnet macaques. Int J Primatol 25:847–859
Singh M, Erinjery JJ, Kavana TS, Roy K, Singh M (2011) Drastic population decline and conservation prospects of roadside dark-bellied bonnet macaques (*Macaca radiata radiata*) of southern India. Primates 52:149–154
Sinha A, Kumar RS, Gama N, Madhusudan MD, Mishra C (2006) Distribution and conservation status of the Arunachal macaque, *Macaca munzala*, in western Arunachal Pradesh, northeastern India. Primate Conserv 21:145–148
Southwick CH, Siddiqi MR (1968) Population trends of rhesus monkeys in villages and towns of northern India, 1959–65. J Anim Ecol 37:199–204
Southwick CH, Siddiqi MF (1988) Partial recovery and a new population estimate of rhesus monkey populations in India. Am J Primatol 16:187–197
Southwick CH, Siddiqi MF (2001) Status, conservation and management of primates in India. ENVIS Bulletin, Wildlife and Protected Areas 1(1): 81–91
Southwick CH, Siddiqi MF (2011) India's rhesus populations: protectionism versus conservation management. In: Gumert MD, Fuentes A, Jones-Engel L (eds) Monkeys on the edge: ecology and management of long-tailed macaques and their interface with humans. Cambridge University Press, Cambridge, pp 275–292
Southwick CH, Beg MA, Siddiqi MR (1961) A population survey of rhesus monkeys in villages, towns and temples of northern India. Ecology 42:538–547

Southwick CH, Siddiqi MF, Oppenheimer JR (1983) Twenty-year changes in rhesus monkey populations in agricultural areas of northern India. Ecology 64:434–439

Southwick CH, Malik I, Siddiqi MF (2005) Rhesus commensalism in India: problems and prospects. In: Paterson JD, Wallis J (eds) Commensalism and conflict: the human–primate interface. American Society of Primatologists, Norman, pp 241–257

Sprague DS (2002) Monkeys in the backyard: encroaching wildlife and rural communities in Japan. In: Fuentes A, Wolfe LD (eds) Primates face to face: the conservation implications of human–nonhuman primate interconnections. Cambridge University Press, Cambridge, pp 254–272

Srivastava A (2006) Conservation of threatened primates of northeast India. Primate Conserv 20:107–113

Srivastava A, Begum F (2005) City monkeys (*Macaca mulatta*): a study of human attitudes. In: Paterson JD, Wallis J (eds) Commensalism and conflict: the human–primate interface. American Society of Primatologists, Norman, pp 259–269

Strum SC (1994) Prospects for management of primate pests. Revue d'Ecologie (La Terre et la Vie) 49:295–306

Supriatna J, Froehlich JW, Erwin JE, Southwick CH (1992) Population, habitat and conservation status of *Macaca maurus*, *Macaca tonkeana* and their putative hybrids. Trop Biodivers 1:31–48

Sussman R, Shaffer CA, Guidi L (2011) Macaca fascicularis in Mauritius: implications for macaque-human interactions and for future research on long-tailed macaques. In: Gumert MD, Fuentes A, Jones-Engel L (eds) Monkeys on the edge: ecology and management of long-tailed macaques and their interface with humans. Cambridge University Press, Cambridge, pp 207–235

Suzuki K, Muroyama Y (2010) Resolution of human–macaque conflicts: changing from top-down to community-based damage management. In: Nakagawa N, Nakamichi M, Sugiura H (eds) The Japanese macaques. Springer, Tokyo, pp 359–373

Thierry B (2011) The macaques: a double-layered society. In: Campbell CJ, Fuentes A, MacKinnon KC, Bearder SK, Stumpf RM (eds) Primates in perspective, 2nd edn. Oxford University Press, Oxford, pp 229–241

van Lavieren E (2008) The illegal trade in Barbary macaques from Morocco and its impact on the wild population. TRAFFIC Bull 21:123–130

van Lavieren E, Wich SA (2010) Decline of the Barbary macaque *Macaca sylvanus* in the cedar forest of the Middle Atlas Mountains, Morocco. Oryx 44:133–138

Wang SW, Curtis PD, Lassoie JP (2006) Farmer perceptions of crop damage by wildlife in Jigme Singye Wangchuck National Park, Bhutan. Wildl Soc Bull 34:359–365

Watanabe K, Muroyama Y (2005) Recent expansion of the range of Japanese macaques, and associated management problems. In: Paterson JD, Wallis J (eds) Commensalism and conflict: the human–primate interface. American Society of Primatologists, Norman, pp 401–419

Wheatley BP (1999) The sacred monkeys of Bali. Waveland Press, Prospect Heights

Wheatley BP (2011) Ethnophoresy: the exotic macaques of Ngeaur Island, Republic of Palau. In: Gumert MD, Fuentes A, Jones-Engel L (eds) Monkeys on the edge: ecology and management of long-tailed macaques and their interface with humans. Cambridge University Press, Cambridge, pp 118–156

Whitten T, Henderson GS, Mustafa M (2002) The ecology of Sulawesi. Periplus, Hong Kong

Wolfe LD (2002) Rhesus macaques: a comparative study of two sites, Jaipur, India, and Silver Springs, Florida. In: Fuentes A, Wolfe LD (eds) Primates face to face: the conservation implications of human–nonhuman primate interconnections. Cambridge University Press, Cambridge, pp 310–330

Woodroffe R, Thirgood S, Rabinowitz A (2005) People and wildlife: conflict or coexistence? Cambridge University Press, Cambridge

Yamada A, Muroyama Y (2010) Effects of vegetation type on habitat use by crop-raiding Japanese macaques during a food-scarce season. Primates 51:159–166

Zhao Q-K (1996) Etho-ecology of Tibetan macaques at Mount Emei, China. In: Fa JE, Lindburg DG (eds) Evolution and ecology of macaque societies. Cambridge University Press, Cambridge, pp 263–289

Zhao Q-K (1999) Responses to seasonal changes in nutrient quality and patchiness of food in a multigroup community of Tibetan macaques at Mt. Emei. Int J Primatol 20:511–524

Zhao Q-K (2005) Tibetan macaques, visitors, and local people at Mt. Emei: problems and countermeasures. In: Paterson JD, Wallis J (eds) Commensalism and conflict: the human–primate interface. American Society of Primatologists, Norman, pp 377–399

Zhao Q-K, Deng Z-Y (1992) Dramatic consequences of food handouts to *Macaca thibetana* at Mount Emei, China. Folia Primatol 58:24–31

Ziegler T, Abegg C, Meijaard E, Perwitasari-Farajallah D, Walter L, Hodges JK, Roos C (2007) Molecular phylogeny and evolutionary history of Southeast Asian macaques forming the *M. silenus* group. Mol Phylogenet Evol 42:807–816

Index

S. Radhakrishna et al. (eds.), *The Macaque Connection: Cooperation and Conflict between Humans and Macaques*, Developments in Primatology: Progress and Prospects 43, DOI 10.1007/978-1-4614-3967-7, © Springer Science+Business Media, LLC 2013

I

J

MIX
Papier aus verantwortungsvollen Quellen
Paper from responsible sources
FSC® C105338

If you have any concerns about our products,
you can contact us on
ProductSafety@springernature.com

In case Publisher is established outside the EU,
the EU authorized representative is:
Springer Nature Customer Service Center GmbH
Europaplatz 3, 69115 Heidelberg, Germany

Printed by Libri Plureos GmbH
in Hamburg, Germany